Applied Ecology and Natural Resource Management

The science of ecology and the practice of management are critical to our understanding of the Earth's ecosystems and our efforts to conserve them. This book attempts to bridge the gap between ecology and natural resource management and, in particular, focuses on the discipline of plant ecology as a foundation for vegetation and wildlife management. It describes how concepts and approaches used by ecologists to study communities and ecosystems can be applied to their management. Guy R. McPherson and Stephen DeStefano emphasize the importance of thoughtfully designed and carefully conducted scientific studies to both the advancement of ecological knowledge and the application of techniques for the management of plant and animal populations. The book is aimed at natural resource managers, as well as graduate and advanced undergraduate students, who are familiar with fundamental ecological principles and who want to use ecological knowledge as a basis for the management of ecosystems.

GUY R. McPHERSON is Professor of Renewable Natural Resources and Ecology and Evolutionary Biology at the University of Arizona in Tucson.

STEPHEN DeSTEFANO is Leader of the U.S. Geological Survey's Massachusetts Cooperative Fish and Wildlife Research Unit, and Adjunct Associate Professor in the Department of Natural Resources Conservation, University of Massachusetts, Amherst.

Applied Ecology and Natural Resource Management

Guy R. McPherson

University of Arizona
School of Renewable Natural Resources and
Department of Ecology and Evolutionary Biology

and

Stephen DeStefano

United States Geological Survey
Massachusetts Cooperative Fish
    and Wildlife Research Unit
University of Massachusetts

CAMBRIDGE
UNIVERSITY PRESS

PUBLISHED BY THE PRESS SYNDICATE OF THE UNIVERSITY OF CAMBRIDGE
The Pitt Building, Trumpington Street, Cambridge, United Kingdom

CAMBRIDGE UNIVERSITY PRESS
The Edinburgh Building, Cambridge CB2 2RU, UK
40 West 20th Street, New York, NY 10011-4211, USA
477 Williamstown Road, Port Melbourne, VIC 3207, Australia
Ruiz de Alarcón 13, 28014 Madrid, Spain
Dock House, The Waterfront, Cape Town 8001, South Africa

http://www.cambridge.org

First published 2003

Printed in the United Kingdom at the University Press, Cambridge

Typeface Swift 9/13pt   *System* QuarkXPress® [TB]

A catalogue record for this book is available from the British Library

Library of Congress Cataloguing in Publication data

McPherson, Guy R. (Guy Randall), 1960–
Applied ecology and natural resource management / Guy R. McPherson and
Stephen DeStefano.
    p.   cm.
Includes bibliographical references (p.).
ISBN 0 521 81127 9 (hb) ISBN 0 521 00975 8 (pbk.)
1. Ecosystem management.   I. DeStefano, Stephen, 1956–   II. Title.
QH75.M3843   2003   333.95–dc21   2002025908

ISBN 0 521 81127 9 hardback
ISBN 0 521 00975 8 paperback

To the managers of natural resources who are dedicated to lifelong learning; may the future rest in their able hands

Contents

Preface

At the risk of merely adding to the bloated and growing literature available on the disciplines of ecology and management while making little meritorious contribution to either, this book attempts to bridge the gap between these literatures and disciplines. As with most books, there are few data and concepts in this text that have not been recorded previously. However, ecology and management have not always been explicitly linked, although each discipline can benefit from the other.

There are many ways that one could link applied ecology to the management of natural resources. Our approach is to focus on plant ecology, and to use this discipline as a foundation for vegetation management. Plant ecology and vegetation management are, in turn, critically important to animal ecology and wildlife management; in many cases, wildlife managers practice vegetation management more directly than they actually "manage" wildlife populations. This additional step – connecting ecologically based vegetation management to wildlife ecology and management – is also frequently recognized but seldom described explicitly, even though it is widely acknowledged that each enterprise can, and does, benefit from the other. Our approach is to use the wealth of information on plant ecology as a basis for the management of both plant and animal populations and natural communities. This book should be especially useful to wildlife ecologists and managers, as it will give insight into the concepts and approaches that plant ecologists use to examine plant communities.

Traditionally, the term "wildlife" has been synonymous with "game," and only species that were hunted were considered worthy of study or management. Some still believe that the fields of wildlife ecology and management are concerned primarily with deer, ducks, and grouse; professional wildlife biologists have moved well beyond this narrow approach. A similar bias might describe the interests of plant ecologists as being limited to pine plantations and row crops. Wildlife ecologists still study species that have recreational or economic importance, but the field of wildlife ecology has evolved. In this book, we define wildlife as any population of vertebrate or invertebrate animals and our interest is in linking our understanding of plant and animal communities to the management of ecosystems. In fact, most ecological principles – and many management practices – that are applicable to a few well-studied species will also apply to many other, lesser known, species. One message that we hope to convey is that it is the questions posed, and the approaches used to address those questions, which are important, rather than the target organism(s) or species of interest.

Many of the concepts and hypotheses within the data-rich disciplines of plant and animal ecology have not been applied to environmental problem-solving. This inability or unwillingness to apply ecological information is vexing and frustrating to scientists who generate knowledge and to managers who attempt to apply that knowledge. The gap between ecological knowledge and application of that knowledge provides the impetus for this book. Thus, this book is designed to organize and evaluate concepts, hypotheses, and data relevant to the application of ecological principles. It serves as a portal into a vast and growing literature on plant and animal ecology and it provides sufficient references to allow the continued exploration of many ecological topics. Most importantly, it provides a framework for the application of the science of ecology to management of ecosystems. The target audience is students and managers who are familiar with fundamental ecological principles and who want to use ecological knowledge as a basis for the management of ecosystems. We are explicitly targeting both students and managers for several reasons. Progressive managers are committed to lifelong learning and are, therefore, students themselves and, as such, this book represents a convenient starting point for new students and an opportunity to refresh, re-evaluate, and "catch up" for managers who have been out of the classroom for some time. Further, the boundaries between the "student" audience and the "manager" audience have eroded, as indicated by the student body in most academic resource-management departments. As recently as 10 years ago, we used the term "nontraditional" to describe students past their 20s; today, these students comprise a significant proportion of most classrooms, and their ranks include many mid-career professionals.

Chapter 1 establishes the foundation for this book and discusses the integration of ecology and management. We begin the chapter with a description of ecology as science. This would seem obvious to some readers, but most of the public in the United States still fails to see ecology as a science and the management of plant and animal populations as an endeavor based on science. One of our goals is to illustrate and promote these relationships and connections. The four chapters that follow address specific topics related to the ecology of plant populations and the implications for animal populations. In Chapter 2, we discuss interactive relationships among organisms – the stuff that makes ecology ecology. Chapter 3 is an in-depth discussion of community structure and a review of techniques that ecologists use to describe structure. In Chapter 4, we address vegetation succession, including a history of concepts, methods to study and manipulate vegetation succession, and the critical role of vegetation succession in shaping communities. In Chapter 5, we close the circle by attempting to narrow the gap between science and management, emphasizing the importance of thoughtfully designed and carefully conducted scientific studies to both the advancement of ecological knowledge and the application of techniques for the management of plant and animal populations.

We have tried to make this book succinct, readable, and affordable. While it is our hope that it is all of these things, our real intention is to assist managers and students in their attempts to connect plant ecology with animal populations, theory with application, and science with management, and to act as a springboard to additional reading and an impetus to the establishment of working relationships between scientists and managers. With respect to the academic student audience, this book is intended to be used as a textbook for graduate or upper-level undergraduate courses in applied ecology. Depending on the specific interests of students and instructors, a course undoubtedly will require supplemental readings, some of which may be referenced herein. For example, an advanced course in applied ecology could supplement this text with a discussion of discriminant analysis and thorough discussion of several of the references in Chapter 3.

Although we have made every effort to make the book palatable reading, there is no question that some of the material it contains is conceptually difficult. For example, the review of models in Chapter 3 is intellectually challenging, particularly for readers new to the concepts. However, this information is fundamentally important to progressive, science-based management. Recalcitrant readers who resist new ideas will not want to read, reflect on, and understand this material; this book is not intended for them.

Figure P.1 Sonoran Desert, Organ Pipe Cactus National Monument, Arizona. Photo by Stephen DeStefano.

The field of ecology continues to grow, and the importance of effective, science-based management of natural resources increases with each passing day. The science of ecology and the practice of management are critical to our understanding of the Earth's ecosystems and our efforts to conserve them (Figure P.1).

ACKNOWLEDGMENTS

Several of our colleagues at the University of Arizona have generously provided moral support and good humor. Bob Steidl (University of Arizona), Jake Weltzin (University of Tennessee), and David Wester (Texas Tech University) supplied ideas, examples, encouragement, and much-needed reviews.

Constructive reviews of parts or all of the manuscript were provided by Cindy Salo, Erika Geiger, Cody Wienk, Heather Schussman, Don Falk, Kristen Widmer, and members of the 1997, 1999, and 2001 versions of the Advanced Applied Plant Ecology class at the University of Arizona. Their efforts greatly improved the manuscript.

Few of the ideas in this book are uniquely ours. We have borrowed them from colleagues, many of whom are mentioned in the preceding paragraphs. We thank them for their insight, and ask their forgiveness for losing track of who had the ideas first. Errors of fact or interpretation remain ours.

Guy McPherson

My wife, Sheila Merrigan, serves as a constant source of inspiration and stability in my life. Neither my career nor this book would have been possible without her.

Many of the ideas in this book can be traced to my mentor and colleague, David Wester. His graduate course at Texas Tech University, Synecology, set a standard by which I gauge my teaching efforts. Wester's course served as the basis for the chapter on community structure. Inspiration and ideas for the chapter on interactions were derived from Paul Keddy's (1989) book, *Competition*. Although we have not met, Keddy has been a role model in my pursuit of scholarship.

This book was derived from notes used to teach a graduate course, Advanced Applied Plant Ecology, at the University of Arizona. I taught the course between 1992 and 2001 to a diverse group of students with majors in natural resource management, ecology, biology, geography, arid land studies, and anthropology. These students have been sufficiently interested in ecology to challenge my knowledge and my teaching style, to the benefit of both. Their interest inspired this text; as such, they share responsibility for its development.

Steve DeStefano

It is not customary to thank one's co-author, but in this case it is appropriate. Guy McPherson provided me with the opportunity to contribute to this book, adding examples and insights of animal biology as they relate to plant ecology. Much of the time and effort that wildlife biologists spend in the field are focused on habitat, and, although vegetation is only one component of habitat, it is an important and often measured variable. My collaboration with Guy has not only allowed me to interject wildlife examples into the book, but has also provided me with the opportunity to learn how plant ecologists think and spend their time. These sorts of collaborations are critical to the advancement of science and its application to resource management, and one of my major hopes for the use of this book is that other wildlife biologists will learn from a premier plant ecologist's perspectives.

I especially thank my friend, colleague, and wife, Kiana Koenen DeStefano. Ki more than anyone encouraged me to realign my priorities, put aside the daily busy work, and "get to work on the book." I also thank her for her insights and the many discussions we have had on wildlife ecology in and out of the field. My life, and the profession of wildlife ecology, are better because of her. I also thank my parents for their constant support and encouragement in all aspects of my life, personal and professional.

Many of the examples in this book drew from my experiences as a field biologist. For those opportunities I thank Drs. Donald H. Rusch, E. Charles Meslow, O. Eugene Maughan, Christopher Brand, and Maurice Hornocker. I also thank the many state and federal agency biologists and managers, university faculty members, and graduate students with whom I have had the pleasure to work.

1

Integrating ecology and management

Ecology is the scientific study of the interactions that determine the distribution and abundance of organisms (Krebs 1972). Predicting and maintaining or altering the distribution and abundance of various organisms are the primary goals of natural resource management; hence, the effective management of natural ecosystems depends on ecological knowledge. Paradoxically, management of ecosystems often ignores relevant ecological theory and many ecological investigations are pursued without appropriate consideration of management implications. This paradox has been recognized by several agencies and institutions (e.g., National Science Foundation, U.S. Forest Service, U.S. Fish and Wildlife Service, Bureau of Land Management, Environmental Protection Agency) (Grumbine 1994; Alpert 1995; Keiter 1995; Brunner and Clark 1997) and entire journals are dedicated to the marriage of ecology and management (e.g., *Journal of Applied Ecology, Conservation Biology, Ecological Applications*). Nonetheless, the underlying causes of this ambiguity have not been determined and no clear prescriptions have been offered to resolve the paradox. The fundamental thesis of this book is that ecological principles can, and should, serve as the primary basis for the management of natural ecosystems, including their plant and animal populations.

Some readers will undoubtedly argue that managers are not interested in hearing about ecologists' problems, and vice versa. Although we fear this may be true, we assume that progressive managers and progressive scientists are interested in understanding problems and contributing to their solution. Indeed, progressive managers ought to be scientists, and progressive scientists ought to be able to assume a manager's perspective. As such, effective managers will understand the hurdles faced by research ecologists, and the trade offs associated with the different methods used to address issues of bias, sample size, and so on. Managers

1

and scientists will be more effective if they understand science and management. How better to seek information, interpret scientific literature, evaluate management programs, or influence research than to understand and appreciate ecology and management?

ECOLOGY AS A SCIENCE

As with any human endeavor, the process of science shares many characteristics with "everyday" activities. For example, observations of recurring events – a fundamental attribute of science – are used to infer general patterns in shopping, cooking, and donning clothing: individuals and institutions rely on their observations and previous experience to make decisions about purchasing items, preparing food, and selecting clothing. This discussion, however, focuses on features that are unique to science. It assumes that science is obliged in part to offer explanatory and predictive power about the natural world. An additional assumption is that the scientific method, which includes explicit hypothesis testing, is the most efficient technique for acquiring reliable knowledge. The scientific method should be used to elucidate mechanisms underlying observed patterns; such elucidation is the key to predicting and understanding natural systems (Levin 1992; but see Pickett *et al.* 1994). In other words, we can observe patterns in nature and ask why a pattern occurs, and then design and conduct experiments to try to answer that question. The answer to the question "why" not only gives us insight into the system in which we are interested, but also gives us direction for the manipulation and management of that resource (Gavin 1989, 1991).

From a modern scientific perspective, a hypothesis is a candidate explanation for a pattern observed in nature (Medawar 1984; Matter and Mannan 1989); that is, a hypothesis is a potential reason for the pattern and it should be testable and falsifiable (Popper 1981). Hypothesis testing is a fundamental attribute of science that is absent from virtually all other human activities. Science is a process by which competing hypotheses are examined, tested, and rejected. Failure to falsify a hypothesis with an appropriately designed test is interpreted as confirmatory evidence that the hypothesis is accurate, although it should be recognized that alternative and perhaps as yet unformulated hypotheses could be better explanations.

A hypothesis is not merely a statement likely to be factual, which is then "tested" by observation (McPherson 2001a). If we accept any statement (e.g., one involving a pattern) as a hypothesis, then the scientific method need not be invoked – we can merely look for the pattern. Such statements are not hypotheses (although the term is frequently applied to them); they are more appropriately called predictions. Indeed, if observation is sufficient to develop reliable knowledge, then science has little to offer beyond everyday activities. Much ecological research is terminated after the discovery of a pattern and the cause of the pattern is not determined (Romesburg 1981; Willson 1981). For example, multiple petitions to list the northern goshawk (*Accipiter gentilis atricapillus*) under the Endangered Species Act of 1978 as a Threatened or Endangered Species in the western United States prompted several studies of their nesting habitat (Kennedy 1997; DeStefano 1998). One pattern that emerged from these studies is that goshawks, across a broad geographical range from southeastern Alaska to the Pacific Northwest to the southwestern United States, often build their nests in forest stands with old-growth characteristics, i.e., stands dominated by large trees and dense cover formed by the canopy of these large trees (Daw *et al.* 1998). This pattern has been verified, and the existence of the pattern is useful information for the conservation and management of this species and its nesting habitat. However, because these studies were observational and not experimental, we do not know *why* goshawks nest in forest stands with this kind of structure. Some likely hypotheses include protection offered by old-growth forests against predators, such as great horned owls (*Bubo virginianus*), or unfavorable weather in secondary forests, such as high ambient temperatures during the summer nesting season. An astute naturalist with sufficient time and energy could have detected and described this pattern, but the scientific method (including hypothesis testing) is required to answer the question of *why*. Knowledge of the pattern increases our information base; knowledge of the mechanism underlying the pattern increases our understanding (Figure 1.1).

Some researchers have questioned the use of null hypothesis testing as a valid approach in science. The crux of the argument is aimed primarily at: (1) the development of trivial or "strawman" null hypotheses that we know a priori will be false; and (2) the selection of an arbitrary α-level or P-value, such as 0.05 (Box 1.1). We encourage readers to peruse and consider the voluminous and growing literature on this topic (e.g., Harlow *et al.* 1997; Cherry 1998; Johnson 1999; Anderson *et al.* 2000). Researchers such as Burnham and Anderson (1998) argue that we should attempt to estimate the magnitude of differences between or among experimental groups (an *estimation problem*) and then decide if these differences are large enough to justify inclusion in a model (a *model selection problem*). Inference would thus be based on multiple model

Figure 1.1 Northern goshawks are often found nesting in stands of older trees, possibly because of the protection offered from predators or weather. Photo by Stephen DeStefano.

building and would use information theoretic techniques, such as Akaike's Information Criterion (AIC) (Burnham and Anderson 1998), as an objective means of selecting models from which to derive estimates and variances of parameters of interest (Box 1.2). In addition, statistical hypothesis testing can, and should, go beyond simple tests of significance at a predetermined P-value, especially when the probability of rejecting the null hypothesis is high. For example, to test the null hypothesis that annual survival rates for male and female mule deer do not differ is to establish a "strawman" hypothesis (D. R. Anderson, personal communication; Harlow *et al.* 1997). Enough is known about the demography of deer to realize that the annual survival of adult females differs from adult males. Thus, rejecting this null hypothesis does not advance our knowledge. In this and many other cases, it is time to advance beyond a simple rejection of the null hypothesis and to seek accurate and precise estimates of parameters of interest (e.g., survival) that will indicate *what* and *how different* the survival rates are for these age-and-sex cohorts. Another approach is to design an experiment rather than an observational study, and to craft more interesting hypotheses: for example, does application of a drug against avian cholera improve survival in snow geese? In this case, determining *how different* would be important, but even a simple rejection of the null hypothesis would be interesting and informative.

Box 1.1 Null model hypothesis testing

The testing of null hypotheses has been a major approach used by ecologists to examine questions about natural systems (Cherry 1998; Anderson *et al.* 2000). Simply stated, null hypotheses are phrased so that the primary question of interest is that there is no difference between two or more populations or among treatment and control groups. The researcher then hopes to find that there is indeed a difference at some prescribed probability level – often $P \leq 0.05$, sometimes $P \leq 0.1$. Criticism of the null hypothesis approach has existed in some scientific fields for a while, but is relatively new to ecology. Recent criticism of null hypothesis testing and the reporting of P-values in ecology has ranged from suggested overuse and abuse to absolute frivolity and nonsensicality, and null hypotheses have been termed strawman hypotheses (i.e., a statement that the scientist knows from the onset is not true) by some authors. Opponents to null hypothesis testing also complain that this approach often confuses the interpretation of data, adds very little to the advancement of knowledge, and is not even a part of the scientific method (Cherry 1998; Johnson 1999; Anderson *et al.* 2000).

Alternatives to the testing of null hypotheses and the reporting of P-values tend to focus on the estimation of parameters of interest and their associated measures of variability. The use of confidence interval estimation or Bayesian inference have been suggested as superior approaches (Cherry 1996). Possibly the most compelling alternative is the use of information theoretic approaches, which use model building and selection, coupled with intimate knowledge of the biological system of interest, to estimate parameters and their variances (Burnham and Anderson 1998). The questions then focus on the values of parameters of interest, confidence in the estimates, and how estimates vary among the populations of interest. Before any of these approaches are practiced, however, the establishment of clear questions and research hypotheses, rather than null hypotheses, is essential.

These arguments against the use of *statistical hypotheses* are compelling and important, but are different, in our view, from the development of *research hypotheses* and the testing of these hypotheses in an *experimental framework*. It is the latter that we suggest is fundamental

Box 1.2 Model selection and inference

Inference from models can take many forms, some of which are misleading. For example, collection of large amounts of data as fodder for multivariate models without a clear purpose can lead to spurious results (Rexstad *et al.* 1988; Anderson *et al.* 2001). A relatively new wave of model selection and inference, however, is based on information theoretic approaches. Burnham and Anderson (1998:1) describe this as "making valid inferences from scientific data when a meaningful analysis depends on a model." This approach is based on the concept that the data, no matter how large the data set, will only support limited inference. Thus, a proper model has: (1) the full support of the data, (2) enough parameters to avoid bias, and (3) not too many parameters (so that precision is not lost). The latter two criteria combine to form the "Principle of Parsimony" (Burnham and Anderson 1992): a trade off between the extremes of underfitting (not enough parameters) and overfitting (too many parameters) the model, given a set of a priori alternative models for the analysis of a given data set.

   One objective method of evaluating a related set of models is "Akaike's Information Criterion" (AIC), based on the pioneering work of mathematician Hirotugu Akaike (Parzen *et al.* 1998). A simplified version of the AIC equation can be written as:

$$AIC = DEV + 2K,$$

where DEV is deviance and K is the number of parameters in the model. As more parameters (structure) are added to the model, the fit will improve. If model selection were based only on this criterion, one would end up always selecting the model with the most possible parameters, which usually results in overfitting, especially with complex data sets. The second component, K, is the number of parameters in the model and serves as a "penalty" in which the penalty increases as the number of parameters increase. AIC thus strikes a balance between overfitting and underfitting. Many software packages now compute AIC. In very general terms, the model with the lowest AIC value is the "best" model, although other approaches such as model averaging can be applied.

   The development of models within this protocol depends on the a priori knowledge of both ecologists and analysts working together, rather than the blind use of packaged computer programs. Information theoretic approaches allow for the flexibility to develop a related set of models, based on empirical data, and to select among or weight those models based on objective criteria. Parameters of interest, such as survival rates or abundance, and their related measures of variance can be computed under a unified framework, thereby giving the researcher confidence that these estimates were determined in an objective manner.

to advancing our knowledge of ecological processes and our ability to apply that knowledge to management problems.

Use of sophisticated technological (e.g., microscopes) or methodological (e.g., statistical) tools does not imply that hypothesis testing is involved, if these tools are used merely to detect a pattern. Pattern recognition (i.e., assessment of statements likely to be factual) often involves significant technological innovation. In contrast, hypothesis testing is a scientific activity that need not involve state-of-the-art technology.

TESTING ECOLOGICAL HYPOTHESES

Some ecologists (exemplified by Peters 1991) have suggested that ecology makes the greatest contribution to solving management problems by developing predictive relationships based on correlations. This view suggests that ecologists should describe as many patterns as possible, without seeking to determine underlying mechanisms. An even more extreme view is described by Weiner (1995), who observed that considerable ecological research is conducted with no regard to determining patterns *or* testing hypotheses. In contrast to these phenomenological viewpoints, most ecologists subscribe to a central tenet of modern philosophy of science: determining the mechanisms underlying observed patterns is fundamental to understanding and predicting ecosystem response, and therefore is necessary for improving management (e.g., Simberloff 1983; Hairston 1989; Keddy 1989; Matter and Mannan 1989; Campbell *et al.* 1991; Levin 1992; Gurevitch and Collins 1994; Weiner 1995; McPherson and Weltzin 2000; McPherson 2001a; but see also Pickett *et al.* 1994).

Since hypotheses are merely candidate explanations for observed patterns, they should be tested. Experimentation (i.e., artificial application of treatment conditions followed by monitoring) is an efficient and appropriate means for testing hypotheses about ecological phenomena; it is also often the only means for doing so (Simberloff 1983; Campbell *et al.* 1991). Experimentation is necessary for disentangling important driving variables which may be correlated strongly with other factors under investigation (Gurevitch and Collins 1994). Identification of the underlying mechanisms of vegetation change enables scientists to predict vegetation responses to changes in variables that may be "driving" or directing the system, such as water, temperature, or soil nutrients. Similarly, understanding the ultimate factors that underlie animal populations will allow wildlife managers to focus limited resources on areas that will likely be most useful in the recovery and management of the population. An appropriately implemented experimental approach yields levels of certainty that are the most useful to resource managers (McPherson and Weltzin 2000).

In contrast to the majority of ecologists, most managers of ecosystems do not understand the importance of experiments in determining mechanisms. In the absence of experimental research, managers and policy-makers must rely on the results of descriptive studies. Unfortunately, these studies often produce conflicting interpretations of underlying mechanisms and are plagued by weak inference (Platt 1964): descriptive studies (including "natural" experiments, *sensu* Diamond 1986) are forced to infer mechanism based on pattern. They are, therefore, poorly suited for determining the underlying mechanisms or causes of patterns because there is no test involved (Popper 1981; Keddy 1989). Even rigorous, long-term monitoring is incapable of revealing causes of change in plant or animal populations because the many factors that potentially contribute to shifts in species composition are confounded (e.g., Wondzell and Ludwig 1995).

Examples of "natural" experiments abound in the ecological literature, but results of these studies should be interpreted judiciously. For example, researchers have routinely compared recently burned (or grazed) areas with adjacent unburned (ungrazed) areas and concluded that observed differences in species composition were the direct result of the disturbance under study. Before reaching this conclusion, it is appropriate to ask why one area burned while the other did not. Preburn differences in productivity, fuel continuity, fuel moisture content, plant phenology, topography, or edaphic factors may have caused the observed fire pattern. Since these factors influence, and are influenced by, species composition, they cannot be ruled out as candidate explanations for postfire differences in species composition (Figure 1.2).

Figure 1.2 Many environmental variables, such as fuel loads, available moisture, and plant phenology, can influence how a fire burns on the landscape. Photo by Guy R. McPherson.

LIMITS TO THE APPLICATION OF ECOLOGY

Considerable research has investigated the structure and function of wildland ecosystems. This research has been instrumental in determining the biogeographical, biogeochemical, environmental, and physiological patterns that characterize these ecosystems. In addition, research has elucidated some of the underlying mechanisms that control patterns of species distribution and abundance. Most importantly, however, research to date has identified many tentative explanations (i.e., hypotheses) for observed ecological phenomena. Many of these hypotheses have not been tested explicitly, which has limited the ability of ecology, as a discipline, to foresee or help solve managerial problems (Underwood 1995). The contribution of science to management is further constrained by the lack of conceptual unity within ecology and the disparity in the goals of science and management.

The unique characteristics of each ecosystem impose significant constraints on the development of parsimonious concepts, principles, and theories. Lack of conceptual unity is widely recognized in ecology (Keddy 1989; Peters 1991; Pickett *et al.* 1994; Likens 1998) and natural resource management (Underwood 1995; Hobbs 1998). The paucity of unifying principles imposes an important dichotomy on science and management: on the one hand, general concepts, which science should strive to attain, have little utility for site-specific management; on the other hand, detailed understanding of a particular system, which is required for effective management, makes little contribution to ecological theory. This disparity in goals is a significant obstacle to relevant discourse between science and management.

In addition, scaling issues may constrain the utility of some scientific approaches (Peterson and Parker 1998). For example, it may be infeasible to evaluate the response to vegetation manipulation of rare or wide-ranging species (e.g., masked bobwhite quail (*Colinus virginianus ridgwayi*), grizzly bear (*Ursus arctos*)). In contrast, common species with small home ranges (e.g, most small mammals) are abundant at relevant spatial and temporal scales and are, therefore, amenable to description and experimentation.

LINKING SCIENCE AND MANAGEMENT

Ecologists have generally failed to conduct experiments relevant to managers (Underwood 1995), and managerial agencies often resist criticisms of performance or suggestions for improvement (Longood and Simmel 1972; Ward and Kassebaum 1972; Underwood 1995). In addition, management agencies often desire immediate answers to management questions, while most ecologists recognize that long-term studies are required to address many questions. These factors have contributed to poorly developed, and sometimes adversarial, relationships between managers and scientists. To address this problem, scientists should be proactive, rather than reactive, with respect to resource management issues, and managers should be familiar with scientific principles. These ideas are developed in further detail in Chapter 5.

Interestingly, some scientists believe that there is insufficient ecological knowledge to make recommendations about the management of natural resources, whereas others believe that ecologists are uniquely qualified to make these recommendations. Of course, decisions about natural resources must be made – the demands of an increasingly large and diverse society necessitate effective management – so it seems appropriate to apply relevant ecological knowledge to these decisions. However, ecologists generally have no expertise in the political, sociological, or managerial aspects of resource management, and they are rarely affected directly by decisions about land management. Thus, ecologists are not necessarily accountable or responsible land stewards. Conversely, managers are ultimately accountable and responsible for their actions, so they should exploit relevant ecological information as one component of the decision-making process. Ultimately, management decisions should be made by managers most familiar with individual systems.

MAKING MANAGEMENT DECISIONS

Management decisions must be temporally, spatially, and objective specific. Thus, management and conservation are ultimately conducted at the local level. Specific management activities, although presumably based on scientific knowledge, are conducted within the context of relevant social, economic, and political issues (*sensu* Brown and MacLeod 1996).

Clearly stated goals and objectives will facilitate management and allow the selection of appropriate tools to accomplish these goals and objectives (Box 1.3). Conversely, selection of goals or objectives that are poorly defined or quantified may actually impede management. For example, use of the term "ecosystem health" implies that there is an optimal state associated with an ecosystem, and that any other state is abnormal; however, the optimal state of an ecosystem must be defined, and clearly stated quantifiable objectives must be developed to achieve that state. Similarly, "ecosystem integrity" (Wicklum and Davies 1995) and sustainability (Lélé and Norgaard 1996) are not objective, quantifiable properties.

Box 1.3 Applying the appropriate fire regime to meet management goals

Throughout the New World, fire regimes changed dramatically after Anglo settlement in concert with changes in ecosystem structure and function. Many ecosystems formerly characterized by frequent, low-intensity surface fires are now characterized by infrequent, high-intensity fires. Altered fire regimes have contributed to, and have resulted from, changes in ecosystem structure; for example, savannas typified by low-intensity surface fires have been replaced in many areas with dense forests that burn infrequently and at high intensities.

Many managers recognize that periodic fires played an important role in the maintenance of ecosystem structure and function, and that these fires probably contributed to high levels of biological diversity. As a result, precise determination of the presettlement fire regime has become an expensive pursuit of many managers. This exercise often is followed by the large-scale reintroduction of recurrent fires into areas where they once were common, in an attempt to restore ecosystem structure by restoring the fire regime.

Unfortunately, accurate reconstruction of events that contributed to historical changes in vegetation (including interruption of fire regimes) will not necessarily facilitate contemporary management, and rarely will engender restoration of presettlement conditions. Pervasive and profound changes have occurred in the biological and physical environments during the last century or more (e.g., dominance of many sites by nonnative species, altered levels of livestock grazing, increased atmospheric CO_2 concentrations). As a result, simply reintroducing periodic fires into areas in which fires formerly occurred will not produce ecosystems that closely resemble those found before Anglo settlement; in this case, understanding the past will *not* ensure that we can predict the future, and a detailed understanding of past conditions may impede contemporary management by lending a false sense of security to predictions based on retrospection. Rather, recurrent fires in these "new" systems may enhance the spread of nonnative species and ultimately cause native biological diversity to decline.

As with any management action, reintroduction of fire should be considered carefully in light of clearly stated, measurable goals and objectives. Historic and prehistoric effects of fires serve as poor analogs for present (and near-future) effects, and presettlement fire regimes should not be used to justify contemporary management. Rather, reintroduction of fires should be evaluated in terms of expected benefits and costs to contemporary management of ecosystems.

The use of terms such as "health," "integrity," and "sustainability" as descriptors of ecosystems implies that managers or scientists can identify the state that is optimum for the ecosystem (vs. optimum for the production of specific resources) and that the preservation of this state is scientifically justifiable. These terms are not supported by empirical evidence or ecological theory, and should be abandoned in favor of other more explicit descriptors (Wicklum and Davies 1995). Appropriate goals and objectives should be identified on a site-specific basis and linked to ecosystem structures or functions that can be defined and quantified.

Pressing needs for the production of some resources and conservation of others indicate that management decisions cannot be postponed until complete scientific information is available on an issue. In

Figure 1.3 Purple loosestrife is a nonnative perennial plant that was introduced into North America in the early 1800s. By the 1930s, it was well established in wetlands and along drainage ditches in the east. Control of this and other exotic species requires consideration of the impact of potential control agents, as well as the nonnative species itself. Photo by Stephen DeStefano.

addition, management goals often change over time. These two considerations dictate the thoughtful implementation of management actions that do not constrain future management approaches and that are targeted at sustaining or increasing biodiversity (e.g., Burton *et al.* 1992). For example, widespread purposeful introduction of nonnative species illustrates a case of near-sighted management focused on the short-term solution of an acute problem, but which reduces future management options by potentially decreasing biodiversity and altering ecosystem structure and function (Abbott and McPherson 1999). Such narrowly focused management efforts are analogous to drilling a hole in the skull of a patient to relieve a severe headache (Figure 1.3).

Like all sciences, ecology is characterized by periodic dramatic changes in concepts. Progressive managers will want to be apprised of these paradigm shifts. For example, the Clementsian model of vegetation dynamics (Clements 1916; Dyksterhuis 1949) still serves as the basis for the classification and management of most public lands, despite the fact that the more appropriate state-and-transition model (Westoby *et al.* 1989) was adopted by ecologists over a decade ago. The delay in adopting the state-and-transition model by land managers probably stems, at least in part, from the absence of an analytical technique to quantify state conditions and transition probabilities (Joyce 1992). The state-and-transition model is described in Chapter 4.

PURSUING RELEVANT ECOLOGICAL KNOWLEDGE

Although descriptive studies are necessary and important for describing ecosystem structure and identifying hypotheses, reliance on this research approach severely constrains the ability of ecology to solve managerial problems. In addition, the poor predictive power of ecology (Peters 1991) indicates that our knowledge of ecosystem function is severely limited (Stanley 1995). An inability to understand ecosystem function and unjustified reliance on descriptive research are among the most important obstacles that prevent ecology from making significant progress toward solving environmental problems and from being a predictive science. Many ecologists (e.g., Hairston 1989; Keddy 1989; Gurevitch and Collins 1994; McPherson and Weltzin 2000) have concluded that field-based manipulative experiments represent a logical approach for future research.

Ecologists can make the greatest contribution to management and conservation by addressing questions that are relevant to resource management and by focusing their research activities at the appropriate temporal and spatial scales (Allen *et al.* 1984). We suggest that these scales are temporally intermediate (i.e., years to decades) and spatially local (e.g., square kilometers), depending on the questions posed and the species of concern. Of course, contemporary ecological research should be conducted within the context of the longer temporal scales and greater spatial scales at which policy decisions are made. For example, experimental research on climate–vegetation interactions should be conducted within individual ecosystems for periods of a few years, but the research should be couched within patterns and processes observed at regional to global spatial scales and decadal to centennial temporal scales. In other words, the context for ecological experiments should be provided by a variety of sources, including observations, management issues (McPherson and Weltzin 2000), long-term databases (Likens 1989; Risser 1991), cross-system comparisons (Cole *et al.* 1991), and large-scale manipulations (Likens 1985; Carpenter *et al.* 1995; Carpenter 1996) (Figure 1.4).

Results of most ecological studies are likely to be highly site specific (Keddy 1989; Tilman 1990) and it is infeasible to conduct experiments in each type of soil and vegetation or for an animal species in every portion

Figure 1.4 Documenting the potential change in geographical distribution of sugar maples and other trees due to global warming requires ecologists to think at large spatial and temporal scales. Photo by Stephen DeStefano.

of its geographical range (Hunter 1989). Therefore, experiments should be designed to have maximum possible generality to other systems (Keddy 1989). For example, the pattern under investigation should be widespread (e.g., shifts in physiognomy), selected species should be "representative" of other species (of similar life form), the factors manipulated in experiments should have broad generality (biomass), experiments should be arranged along naturally occurring gradients (soil moisture, elevation), and experiments should be conducted at spatial (community) and temporal (annual or decadal) scales appropriate to the management of communities.

Ecological experiments need not be conducted at small spatial scales. For example, ecosystem-level experiments (i.e., relatively large-scale manipulation of ecosystems) represent an important, often-overlooked technique that can be used to increase predictive power and credibility in ecology. Ecosystem-level experiments may be used to bridge gaps between small-scale experiments and uncontrolled observations, including "natural" experiments. However, they are difficult to implement and interpret (Carpenter *et al.* 1995; Lawton 1995): they require knowledge of species' natural histories, natural disturbances, and considerable foresight and planning. Fortunately, ecology has generated considerable information about the natural history of dominant species and natural disturbances in many ecosystems. Similarly, foresight and planning should not be limiting factors in scientific research. Time and money will continue to be in short supply, but this situation will grow more serious if ecology does not establish itself as a source of reliable knowledge about environmental management (Peters 1991; Underwood 1995).

In addition to posing questions that are relevant to resource management and that investigate mechanisms, scientists should be concerned with the development of research questions that are tractable. Asking why certain species are present at a particular place and time forces the investigator to rely on correlation. In contrast, asking why species are *not* present (e.g., in locations that appear suitable) forces the investigator to search for constraints, and therefore mechanisms (e.g., DeStefano and McCloskey 1997). Although Harper (1977, 1982) presented a compelling case for tractable, mechanistic research focused on applied ecological issues two decades ago, the underwhelming response by ecologists indicates that his message bears repeating.

SUMMARY

Management decisions must be temporally, spatially, and objective specific, so that management and conservation are ultimately conducted at the local level. Appropriate management can be prescribed only after goals and objectives are clearly defined. After goals and objectives are identified, ecological principles can be used as a foundation for the progressive, effective management of natural resources. Managers of natural resources must be able to distinguish candidate explanations from tested hypotheses, and therefore distinguish between conjecture and reliable knowledge. Ecologists can contribute to management efforts by addressing tractable questions that are relevant to resource management, and by focusing their research activities at appropriate temporal and spatial scales. The following chapters illustrate that the science of ecology can be linked with the management of natural resources in ways that are conducive to the improvement of both endeavors.

2

Interactions

Understanding interactions is fundamental to predicting the distribution and abundance of organisms at spatial and temporal scales appropriate to management. Therefore, this chapter focuses on interactions that are particularly relevant to the management of plant and animal populations. Any or all interactions may assume considerable importance in structuring ecosystems. In general, however, few factors exert primary control over community structure at a specific place and time. Identifying which factor (or factors) primarily affects community structure, particularly in a site- and time-specific manner, is therefore a necessary first step for effective management.

Abiotic factors, such as soil type, hydrology, or weather, assume increasing importance as spatial scales increase beyond the local level and as temporal scales exceed decadal time frames (Prentice 1986; Archer 1993, 1995a). Some abiotic constraints can be overcome with appropriate management, and these are described in the following chapter. This chapter will focus on the methods used to study biotic interactions (i.e., among organisms), discuss the interactions that frequently underlie community structure, and describe techniques that may be used to alter the outcome of interactions (Figure 2.1).

CLASSIFYING INTERACTIONS

Many introductory ecology texts use a conceptually simple strategy to classify interactions (Table 2.1). Five interactions are commonly recognized: competition (mutually detrimental), amensalism (detrimental to one participant, no effect on the other), commensalism (beneficial to one participant, no effect on the other), mutualism or symbiosis (mutually beneficial), and contramensalism (detrimental to one participant, beneficial to the other). Predation, parasitism, and herbivory represent

17

Table 2.1 $+/-$ *system for classifying interactions*

	+	0	−
+	mutualism	commensalism	contramensalism
0	commensalism	no interaction	amensalism
−	contramensalism	amensalism	competition

Figure 2.1 Interactions among plants and wildlife are as varied as biodiversity itself. Plants, vertebrate animals, and invertebrate animals interact to cause patterns of distribution and abundance, and therefore influence the structure of ecosystems. Photo by Stephen DeStefano.

examples of contramensalism (Arthur and Mitchell 1989), and allelopathy is viewed as an extreme form of amensalism.

In practice, many studies are one sided (strongly asymmetrical, *sensu* Weiner 1990): they are designed to assess the impact of only one participant (usually the dominant organism on a site). Thus, the research-based knowledge about interactions often does not parallel the terminology used to describe interactions. For example, much research ostensibly focused on competition involves the manipulation of one participant; as a result, many authors improperly use the term "competition" to describe the detrimental impact of one participant on another (Keddy 1989). "Interference" is a preferred term for describing these interactions unless *mutually detrimental* effects are demonstrated (Harper 1977). Similarly, "facilitation" is preferred over "mutualism" or "symbiosis" if only one participant in the interaction is studied. In practice, it may be impossible to identify every interacting participant: interactions occur that cannot be detected or even surmised. Thus, the use of terms such as competition, amensalism, commensalism, mutualism, and contramensalism should be restricted to cases in which the roles of all the participants in the interaction have been documented. In other cases, more general terms are preferred.

Assumptions have also been made about trophic-level interactions in a community. For example, competition is assumed to operate *within* trophic levels (e.g., among insectivorous birds, seed-eating rodents, or mammalian carnivores), while predation is assumed to operate primarily *between* trophic levels (e.g., carnivores preying on herbivores, herbivores preying on plants). However, some species may act as both competitor and predator with other species within a trophic level (Stapp 1997). This phenomenon, known as *interguild predation* (Polis and McCormick 1986), has been studied in invertebrates but may also be important among vertebrates (Cortwright 1988; Polis *et al.* 1989; Gustafson 1993; Lindström *et al.* 1995; Olson *et al.* 1995; Stapp 1997).

The use of a simple matrix is a useful starting point for a discussion of interactions. However, it must be recognized that reality is more complex than this simple conceptual model. For example, it may not be possible to distinguish realistically amensalism from strongly asymmetric competition. In addition, the participants in an interaction may change roles over time, so that the interaction changes from one category to another. For example, bur-sage (*Ambrosia deltoidea*) initially acts as a "nurse plant" for several species in the Sonoran Desert (McAuliffe 1988). However, this initially commensalistic interaction may become competitive and eventually amensalistic as the plants established in the shade of bursage grow through the canopy of the nurse plant and eventually overtop it. Thus, the relationship between individual plants changes over time (Figure 2.2). Similarly, research with short-lived plants indicates that the symmetry of interactions at the level of populations may change in 20–40 generations (Aarssen and Turkington 1985; Aarssen 1989, 1992; Turkington and Jolliffe 1996). These examples illustrate that the category assigned to an interaction should be viewed in its appropriate context: understanding the nature of the interaction is more important than properly classifying the interaction (Bronstein 1994).

In addition to being overly simplistic, the matrix approach to the study of interactions may be misleading. For example, herbivory $(+, -)$ may or may not be detrimental to one participant, even at the level of the individual plant: the response of a plant to herbivory is strongly

Figure 2.2 Certain species of cactus, such as saquaros, germinate and grow primarily in the protective shadows of other plants, such as bur-sage and palo verde, before they eventually overtop their nurse plants. Photo by Stephen DeStefano.

dependent on plant size and phenology (e.g., Briske and Richards 1995; Evans and Seastedt 1995). In fact, herbivory apparently increases the reproductive output of some plants, which presumably is beneficial to the grazed plant (Whitham *et al.* 1991). Thus, herbivory may be classified as "contramensalism" for some species or individuals, and as "commensalism" or "mutualism" for others. To address this point, we have assumed that herbivory is beneficial for herbivores, which is not necessarily accurate (e.g., in the case of poisonous plants).

The outcome of a specific interaction, within an evolutionary context, provides additional justification for avoiding the matrix approach to classify interactions. Herbivory from native herbivores is rarely sufficient to cause the extinction of herbs (likewise, predation rarely drives prey to extinction). Thus, from an evolutionary standpoint, herbs are still successful (i.e., they are still present), and the interaction can hardly be classified as "detrimental" to herbs. A notable exception involves species that lack a shared evolutionary history; nonnative herbivores or predators, for example, may cause extinction of herbs or prey, respectively, because the organism being eaten evolved in isolation from the organism that is eating it (Box 2.1).

Box 2.1 Free-ranging domestic cats

Domestic cats (*Felis catus*) are effective, efficient, and tireless hunters. Whether they are truly feral (i.e., a domestic animal surviving on its own) or a pet that the owner lets loose, free-ranging cats kill large numbers of songbirds, small mammals, and lizards. Free-ranging domestic cats have been implicated in local extinctions of some populations of songbirds and small mammals (Crooks and Soulé 1999), and they compete with native predators and may reduce their numbers (see Coleman and Temple 1993 for review). Unlike natural predators, whose numbers, reproductive output, and survival depend on adequate populations of prey, numbers of domestic cats are kept artificially high by supplemental feeding. Estimates of cat density range as high as 40–44 cats/km^2 (Coleman and Temple 1993). Domestic cats will also continue to capture prey even while being fed by their owners (Adamec 1976). Despite these concerns, many cat owners continue to insist that their pets be allowed to roam free. Many of these people are also nature lovers and are concerned with wildlife populations, but the attitude that *their* cat would not kill small animals allows this contradictory behavior to exist.

In a recent study in Florida, Castillo (2001) examined what some called "managed" colonies of stray and feral cats. Cats in these colonies are kept fed by people, with the idea being that a well-fed cat will not hunt and kill wildlife. Proponents of cat colonies also believe that cats are territorial, and that their territorial behavior will prevent more cats from joining the colony, and that cat colonies will decline in size over time. Castillo's findings were just the opposite: well-fed cats continued to kill wildlife, and aggressive interactions among cats were few and did not limit the size of the colony. Further, cat feeders attracted other animals, such as skunks, raccoons, and stray dogs, and cat colonies only served to encourage cat abandonment. Some conservation groups, such as the American Bird Conservancy (www.abcbirds.org) have launched campaigns to encourage cat owners to keep their cats indoors.

The ecological literature is replete with studies of interactions – even full-time researchers cannot keep up with the explosively expanding literature. Published papers must be evaluated quickly with respect to their potential relevance and utility. This section classifies various studies into one of four categories – descriptive studies, comparative studies, models, and experiments – and summarizes several detailed comparisons of these approaches (Diamond 1986; Keddy 1989; McPherson and Weltzin 2000).

Descriptive studies

Descriptive research was the traditional approach until the early 1960s, when Connell (1961) and Paine (1963) published seminal experimental papers. Descriptive studies remain widely used, at least partly because of historical precedence: "generations of plant ecologists have been occupied with tallying the contents of quadrats in the summer, and then trying to draw inferences about these observations in the winter" (Keddy 1989:83).

An impressive number of statistical techniques has been developed just for investigating patterns in data sets derived from field descriptions. The biggest problem with the descriptive approach is that a mechanism (i.e., an interaction) must be invoked to explain a pattern, but that several different processes may produce the same pattern. Consider the following simple example, using association analysis (Keddy 1989:83–85). Data are collected from sample units (usually quadrats) and the association between any pair of species is calculated using the chi-square test. The null hypothesis is that the species are independently distributed; the alternative hypothesis is that the two species are either positively or negatively associated. Negative associations are often interpreted as providing evidence of competition, when actually at least four interpretations are possible:

1. Species are restricted to different microhabitats, and so do not interact. For example, the species may possess different physiologies, either as adults or, less conspicuously, as juveniles or seeds.

2. Agents such as predators independently control each species and restrict each to a different set of conditions.
3. Species are positively associated but the sample unit (e.g., quadrat) is so small that only a few individuals fit within it. Thus, the pattern observed at the local scale obscures the pattern at the more relevant larger scale.
4. The species compete, and competition leads to habitat segregation.

In this example, it is not possible to distinguish between these four interpretations with descriptive data alone. In fact, the first two hypotheses cannot be falsified; the inability to find differences between species indicates either that the researcher has not investigated in sufficient depth or that the two taxa are not different species.

A variation of association analysis is to choose natural environmental gradients (e.g., lakeshores, mountains) and then to examine the distributional limits of species along these gradients. Three alternatives are widely recognized (Keddy 1989): (1) species distributional limits are regularly spaced; (2) species distributional limits are randomly arranged; and (3) species distributional limits are clustered along the gradient. Statistical tests have been developed which describe distributions (Pielou 1977; Underwood 1978; Pielou 1979; Shipley and Keddy 1987).

As with association analysis, process is inferred from pattern: systems structured by interactions are assumed to have different kinds of patterns than those not structured by interactions. However, as with association analysis, the relationship between interactions and resulting patterns is not clear. In particular, departures from random patterns do not reveal the presence of an interaction. For example, it has been widely proposed that clumped or regular patterns result from competitive interactions. In fact, at least four interpretations can account for clumped distributional limits (Keddy 1989:86–7):

1. Species have similar distributional limits because of similar physiological tolerance limits. For example, all species possess a similar mechanism to tolerate flooding, so they occupy similar positions along gradients of soil moisture or drainage.
2. Clusters of distributional limits may be attributed to the manner in which the observer divided the gradient.
3. Herbivores may stop at a certain point along a gradient, thereby creating discontinuities in plant distribution.
4. One or more competitive dominants may control the distributional limits of an entire group of species (*sensu* Keddy 1990).

Only the last two hypotheses are consistent with the existence of interactions, and they suggest two different interactions (herbivory, interference). These hypotheses are very difficult to falsify with descriptive data because the existence of new, unexplored environmental gradients or factors can always be postulated.

An equally meaningless set of interpretations could be developed for regular or random patterns, much like the set associated with clumped patterns, with relatively little effort. Despite the inability to distinguish between hypotheses with descriptive data, these data continue to serve as a source of entertainment for ecologists. Two case studies illustrate the application of the descriptive approach to the study of interactions.

Case study: distribution patterns of desert plants

Yeaton *et al.* (1977) used nearest-neighbor analysis to describe patterns of plant distribution in southern Arizona. They correlated the distance between two plants with the sum of the sizes of the plants, and found that: (1) all intraspecific comparisons were significantly correlated; (2) creosotebush (*Larrea tridentata*) was negatively correlated with all other species except saguaro (*Carnegia gigantea*); and (3) bur-sage (*A. deltoidea*) was negatively correlated only with bur-sage and creosotebush, and was positively correlated with saguaro. The latter finding is consistent with bur-sage as a nurse plant for many succulent species in the Sonoran desert (McAuliffe 1988).

Yeaton *et al.* subsequently attempted to correlate patterns of above- and below-ground morphology with the seasonal growth patterns of plants. This aspect of the paper was characterized by strongly stated conclusions based on little evidence.

Case study: spacing of acorn woodpeckers

Campbell (1995) reanalyzed the data of Burgess *et al.* (1982) on the spacing patterns of acorn woodpeckers (*Melanerpes formicivorus*). Based on graphical analysis, Campbell determined that acorn woodpeckers exhibited regular spacing. According to Campbell (1995:136), this spatial pattern "provides good evidence of competition."

The study by Burgess *et al.* (1982) provoked controversy in the ecological literature (Burgess 1983; Mumme *et al.* 1983; Brewer and McCann 1985; Krebs 1989:167–8). Unfortunately, Campbell's reanalysis of these data is unlikely to resolve the controversy: invoking the process of competition from an observed pattern is unlikely to produce a consensus.

Comparative studies

Comparative studies follow directly from descriptive studies, in that observational data are used to describe patterns and the resulting patterns are compared in order to infer differences in process. As with descriptive studies, caution is warranted; in the absence of an experiment, the researcher is forced to compare patterns and then invoke mechanisms. When many hypotheses make similar predictions about pattern, this process simply does not work. Given the complexity of nature, many alternative hypotheses typically can be generated (Keddy 1989).

The primary value in comparative studies lies in the spatial and temporal scales that can be considered. Thus, comparative studies are commonly used to infer the presence of interactions when experimental studies are difficult to conduct (Keddy 1989). The following case studies employed comparative studies because experiments are difficult to conduct with large, mobile animals (birds on islands) and with slow-growing species (lichens).

Case study: birds on islands

Diamond (1975) described the distributions of birds on New Guinea and nearby islands, and used islands as the sample units. Diamond used the groups of species on each island to infer factors that may explain the different assemblages (e.g., dispersal abilities, competition). These observations were augmented by descriptions of habitat and the food requirements of many species.

Diamond's assumption of competition as an important factor structuring bird assemblages was critically challenged by Connor and Simberloff (1979). The ensuing debate, which focused partially on the identification of structures against which to compare observed species distributions (null models), has been heated and divisive (e.g., Grant and Abbott 1980; Diamond and Gilpin 1982; Gilpin and Diamond 1982; Wright and Biehl 1982; Simberloff 1983, 1984). The participants in the debate have ignored the extensive literature on pattern analysis in plant communities, including the construction of null models (e.g., Pielou 1977; Underwood 1978; Pielou 1979; Dale 1984). More importantly, they have not focused attention on the critical issue: the cause(s) of patterns cannot be determined from further studies of pattern, regardless of analytical sophistication (Keddy 1989).

Case study: lichens

Pentecost (1980) compared lichen communities at two sites in the British Isles. Both sites were characterized by two dominant species of crustose lichens. Pentecost determined which (if any) species grew over the other at each point of contact. At the site dominated by closely related species (*Caloplaca heppiana* and *C. aurantia*), individuals typically did not over-grow each other – rather, growth of both individuals usually ceased at the point of contact. Pentecost concluded "that the competitors are well matched at this site" (1980:136). At the second site, *Aspicilia calcarea* usually grew over *C. heppiana* at points of marginal contact, but *C. heppiana* established in windows within *A. calcarea*. Pentecost concluded that competitive interactions between these and other lichen species contribute to patterns of succession.

Models

The use of ecological models requires clear statements of assumptions and rigorous application of methods, such as goodness-of-fit testing, to test the validity of those assumptions, given a set of data. Models can then contribute mathematical rigor and insight to ecological concepts. Models may also influence both the kinds of questions that are asked and the manner in which they are addressed, with consequent impacts on the types of studies that are conducted.

However, some models make little contribution to the development of ecological theory, and even less contribution to the application of theory to the solution of relevant environmental problems. Simberloff (1983:630) noted "... that ecological modeling for its own sake is now a recognized discipline is witnessed by the emergence of journals ... devoted primarily or solely to modeling and scarcely at all to whether the models correspond to nature." Consider, for example, the most commonly studied models of interactions between species: the Lotka – Volterra models. As discussed by Keddy (1989), these models have virtually no resemblance to reality and they are essentially impossible to use in real ecosystems; in addition, the models are not mechanistic. The Lotka–Volterra equations continue to be studied, but they are no longer used to understand real ecosystems (Keddy 1989). In sharp contrast to simple, nonmechanistic Lotka–Volterra models, two complex, mechanistic models of interactions have been developed by Tilman (1982, 1988). These models were designed to explore questions of coexistence in plant communities. Unfortunately, an overwhelming amount of information is needed to construct and parameterize the models. Tilman (1982) has described the requirements for the simpler of the two models: "it will be necessary to know the resource requirements and competitive interactions of the dominant species under controlled conditions, the correlations between the distributions of the species in the field and the distributions of limiting resources, and the effects of various enrichments (of resources, e.g., via fertilization) on the species composition of natural communities."

Modeling in ecology has, however, entered a new phase. Many researchers are recognizing that mathematical models must be based in reality, have a foundation anchored in biology, and be the product of modelers and biologists working closely together (DeStefano *et al.* 1995). Understanding the dynamics of populations and related ecological and evolutionary issues frequently depends on a direct analysis of life history parameters, such as survival (Lebreton *et al.* 1992). These complicated life-history processes are often best examined through a process of modeling that incorporates the construction of multiple models based on real data sets, goodness-of-fit testing to evaluate the appropriateness of the models to a specific set of data, and model selection that incorporates objective criteria or model averaging (Anderson and Burnham 1992; Burnham and Anderson 1998). The strength of these models lies in their ability to provide a unified approach to the estimation of population parameters and their associated measures of variability. Examples include the estimation of density and abundance (Buckland *et al.* 1993) and survival rates (Lebreton *et al.* 1992). Accurate estimates of these types of important population characteristics are critical to furthering our understanding of life history processes, ecology, and conservation and management. Biologists are also recognizing the limitations of models and are becoming increasingly realistic in their use of models to address management problems (e.g., Beissinger and Westphal 1998). Ecological models need not always rely on the language of mathematics, and ecologists should embrace the inherent complexity of ecology, not gloss over it with simplistic mathematical constructs. Thus, we join a growing number of ecologists who discourage the pursuit of mathematical models that do not have a biological basis (e.g., Pielou 1981; Simberloff 1983; Hall 1988, 1991; Grimm 1994), and we encourage the development and use of conceptual and mathematical models which provide a framework for studying actual populations and ecosystems.

Case study: competitive hierarchy model

The competitive hierarchy model (Keddy 1990) is a conceptual model that attempts to explain how species partition resources. It includes three primary assumptions (Keddy 1989): (1) species in the community have inclusive niches (i.e., all species achieve the greatest abundance or growth rate at the same end of the gradient); (2) species vary in competitive ability in a predictable manner, and competitive ability is an inherent trait of a species (this is equivalent to assuming that environment exerts no influence over competitive ability); and (3) competitive abilities are negatively correlated with fundamental niche width (perhaps due to an inherent trade off between ability to interfere with neighbors and ability to tolerate low resource levels).

Relaxing the assumptions of the competitive hierarchy model greatly increases its application and utility. Specifically, consider the assumption that the dominant species occupies the "preferred" end of a resource gradient, and subordinates are displaced down the gradient a distance directly related to their position in the competitive hierarchy (i.e., the realized niche is equivalent to the fundamental niche for the dominant species in any interaction: interference is strictly asymmetric). Relaxing this assumption indicates that there is some point at which the dominant is so weakened by environmental effects that it can be excluded by the subordinate. Depending on how far this assumption is relaxed, a series of cases can be produced that are intermediate between strict resource partitioning and competitive hierarchies (Keddy 1989). In other words, interference affects both ends of a species' distribution (major role at one end, minor role at the other), and environmental influences on competitive interactions are acknowledged.

The competitive hierarchy model has been extended to larger levels of organization by proposing the centrifugal organization model of community structure (Keddy 1990; Keddy and MacLellan 1990; Wisheu and Keddy 1992). This model is discussed in the following chapter.

Case study: resource partitioning among African ungulates

There are more than 15 sympatric ungulate species in African savannas. Although there is evidence that these populations are food limited (Sinclair 1975, 1977), experimental confirmation of competition is lacking (Owen-Smith 1989). Owen-Smith and Novellie (1982) developed an optimal foraging model for large herbivores, and Owen-Smith (1989) used this model to examine the mechanisms that may lead to resource partitioning among ungulates and to investigate the circumstances in which competition is most likely to occur.

The model incorporated an array of potential food types, defined by their nutrient concentrations and abundances, and morphological differences among ungulates, such as digestive anatomy, body mass, and relative mouth dimensions. Mouth dimensions influence bite size, as well as other morphological, physiological, and behavioral components. Food types and morphological variables were used in the model to simulate how food resources may be used among different ungulate species.

The primary findings of this study were: (1) larger ungulates were dependent on lower quality but more abundant food types not eaten by smaller ungulates; (2) nonruminants ate a wider range of food types in terms of quality than did ruminants (the former had a lower digestive efficiency but could exploit a wider range of vegetation types); and (3) differences in mouth size (and thus body size) led to differences in plant species preferences, such as leaf size or height above ground of grasses.

Owen-Smith concluded that differences in morphology and physiology among sympatric ungulate species lead them to favor different vegetation types for foraging, so that resource partitioning may play a larger role in shaping the savanna ungulate community than interspecific competition. He noted that competition may be most likely to occur over uncommon plants that are of exceptionally high nutritional value, and/or during the dry season when forage availability is more restricted – although differences in habitat or dietary choices among ungulates seemed to be more apparent during this time of year. When discussing the potential for competition in this community of herbivores, Owen-Smith (1989:161) stated:

> Mere overlap in resource use cannot be equated with competition. For a competitive relation to exist, the effects of feeding by one species must be such as to reduce the foraging efficiency of another to the detriment of the population density or recruitment of the latter (MacNally 1983).

Experiments

The use of experiments to study interactions was proposed by a prominent ecologist over 80 years ago (Tansley 1914): "in order to determine the powers of the different species, we must resort to experiment." Frederic Clements implemented a classic series of removal experiments shortly thereafter (Clements *et al.* 1929). Considering the influence of Tansley and Clements on the early development of ecology, it is surprising that an experimental approach to the study of interactions was largely ignored until the 1960s (Keddy 1989; Aarssen and Epp 1990).

An experiment requires the researcher to specify a question and a means of answering the question in advance. As a result, experimental studies tend to be better designed than descriptive or comparative studies.

In fact, Keddy (1989) warns against the temptation to collect descriptive or comparative data without first specifying a question: "Because ecologists usually love working with real ecosystems, it is always tempting to pick up quadrats, binoculars, sample bottles or nets and head to field without any questions at all." It is perfectly appropriate to "head to the field without any questions at all," and many ecologists would benefit from spending more time in the field; however, astute ecologists will leave the data forms at home until appropriate questions have been developed.

Descriptive studies and comparative studies do not test for interactions. Rather, they search for a pattern and infer a mechanism. Experiments do not necessarily overcome this problem; they usually eliminate the largest number of alternative hypotheses, but experimental manipulations ultimately test for density dependence. For example, reduction in the abundance of one species (via manipulation) may lead to an increase in abundance of another species, but this relationship does not conclusively demonstrate the existence of a negative interaction between the two species. "Apparent" competition (Connell 1990), resulting from indirect effects, has been proposed as an alternative mechanism for density dependence.

"Apparent" competition necessarily involves complex interactions, but identification of the interaction is not straightforward. For example, a species that is reduced in abundance may have been a host for a pathogen which also damaged the remaining species. In this case, removal of one species appears to release the other species from interference, but instead actually releases it from the effects of the pathogen. Alternatively, the "removed" species may have attracted a herbivore which also fed on the remaining species; thus, removal of the former species releases the latter from herbivory. Parker and Root (1981) demonstrated that a herbaceous plant species was excluded from some habitats by a grasshopper. The grasshopper was typically associated with a common shrub, and removal of the shrub contributed to increased abundance of the herb without any evidence of interaction between the shrub and the herb. These examples indicate that the mechanisms of interaction between species may be very complex, making it difficult to separate direct from indirect effects. They further illustrate that experiments cannot be divorced from natural history (Keddy 1989).

A common conjecture underlying experimental studies is that interactions are structuring a community in a certain manner. A manipulation is then conducted to evaluate the presence and strength of the presumed interaction. Variables are specified in advance, and include abundance of neighbors (independent variables) and some metric(s) of the performance of individuals or populations (dependent variables – e.g., size, change in size, distribution along a gradient). An inverse relationship between independent and dependent variables indicates the occurrence of a detrimental effect of one individual or population on the other (i.e., interference), whereas a direct relationship indicates the occurrence of a beneficial effect (i.e., facilitation). The strength of the relationship indicates the magnitude of the effect. Reciprocal relationships are rarely studied; scientists tend to evaluate the effects of species that appear to be dominant, thereby ignoring potentially important "subordinate" species.

Experimental manipulations may involve the removal or addition of individuals, and they may be conducted in the laboratory or in the field. The relative merits of various experimental designs have been extensively reviewed (Bender *et al.* 1984; Keddy 1989; Campbell *et al.* 1991; Snaydon 1991; Sackville Hamilton 1994; Snaydon 1994). The following sections describe and compare different types of experiments, and provide an example of each.

Removal experiments

Removal experiments are used to determine whether reductions in one species will cause changes in the distribution or abundance of other species. Thus, they may provide evidence that interactions are currently structuring the community. If reduction or removal of a species produces increases in another species, the former species was presumably interfering with the latter. Conversely, if reduction or removal of a species causes another species to decrease in distribution or abundance, the former species was apparently facilitating the latter.

Removal experiments are relatively common in the wildlife and animal ecology literature. Many of these studies involve the reduction of predator populations to benefit prey or game populations, such as ungulates (Boertje *et al.* 1996) or livestock (Conner *et al.* 1998), or to improve survival and reproductive output and thus assist in the recovery of rare or endangered species. Removal experiments in some systems have shown that common predators with generalist diets have reduced species richness or have caused the local extinction of rare species. Examples include avian predators and grasshopper prey (Joern 1986), rodent predators and beetle prey (Parmenter and MacMahon 1988), and lizard predators and spider prey (Spiller and Schoener 1998). These studies have important implications for the management and recovery of threatened or endangered species.

Petren and Case (1996) studied competition between two species of geckos in urban and suburban environments in the Pacific basin. The native species, *Lepidodactylus lugubris*, declines numerically when a non-native species, *Hemidactylus frenatus*, invades its habitat. Replacement of *Lepidodactylus* by *Hemidactylus* occurs rapidly and is facilitated by clumped insect resources, suggesting that the mechanism of displacement is due to the ability of each species to exploit food resources. The fact that these two species show nearly complete diet overlap, that there is no evidence of direct antagonistic behavior between the species, that demographic studies have shown insects to be a limiting resource, and that reduced food resources negatively affect body condition, fecundity, and survival of *Lepidodactylus* supports this hypothesis.

Petren and Case constructed enclosures and developed four replicates of four treatments: *Lepidodactylus* at low density without *Hemidactylus*; *Lepidodactylus* at low density with *Hemidactylus*; *Lepidodactylus* at high density without *Hemidactylus*; and *Lepidodactylus* at high density with *Hemidactylus*. Thus, in effect, the invader *Hemidactylus* was "removed" and no longer coexisted with the native *Lepidodactylus* in two of the four types of enclosures. From their observations of foraging by the captive geckos and its impact on insect abundance within the enclosures as well as body condition, relative foraging ability, and demographic performance of the geckos, the authors concluded that clumped resources increased interspecific competition between *Lepidodactylus* and *Hemidactylus* and that this competition contributed to a change in the gecko species composition, favoring *Hemidactylus*. They attributed the advantage of *Hemidactylus* to its larger body size, faster running speed, and reduced intraspecific interference while foraging. Human alteration of the environment – in this case, the prevalence of lights in urban and suburban environments which increased the clumped distribution of insects – facilitated the competitive advantage of *Hemidactylus* over *Lepidodactylus*.

Petren and Case point out that experiments that measure competition between long coexisting species may underestimate the role of competition in structuring communities. By studying the impact of nonnative species on native competitors, however, we may be able to understand better the role that competition could play in altering community composition.

In many cases, removal of predators is an effective device for increasing the density or reproduction of prey populations. A study conducted in

Australia has demonstrated the dramatic increase in a population of prey that may follow predator removal. In this case, however, the implications for the conservation of native fauna and flora were not straightforward.

Australian natural resource managers are interested in killing nonnative foxes (*Vulpes vulpes*) because the foxes are major predators of native wildlife and domestic lambs. They are also important predators of nonnative wildlife, especially rabbits (*Oryctolagus cuniculus*). Banks *et al.* (1998) used poison on two treatment plots over an 18-month period to eradicate nonnative red foxes. On two control plots, foxes were present but were not poisoned. Rabbit populations on the two treated plots grew to 6–12 times their initial density within the 18 months, while numbers on the two untreated plots showed very small increases over the same period. Because this was a controlled experiment, the researchers correctly concluded that predation affected the rabbit population: experimental removal of foxes led to an impressive increase in the rabbit population. The conservation conundrum, of course, was in the implications of altering the contramensalistic relationship of rabbit and fox: "[A]s fox removal was initially planned to protect native fauna threatened by fox predation, the response of the rabbits represents a serious ecological cost of fox control" (Banks *et al.* 1998:766). Thus, the researchers were able to determine that fox predation was at least one mechanism that controlled rabbit populations, but they also traded one environmental problem for another: intense predation on native fauna by introduced red foxes was replaced with intense herbivory of native flora by introduced rabbits (Figure 2.3).

CASE STUDY: REMOVAL EXPERIMENT – COMPETITION OR PREDATION?
As mentioned previously, competition between or among species can be confused with other types of interactions (e.g., interguild predation). Stapp (1997) addressed this concept in a removal study of shortgrass prairie rodents in shrub-dominated shortgrass prairie in Colorado, asking if the structure of the rodent community was the result of competition or predation. During one summer, Stapp compared abundance, microhabitat use, and diet of deer mice (*Peromyscus maniculatus*) on four areas where northern grasshopper mice (*Onychomys leucogaster*) and deer mice coexisted with four areas where grasshopper mice had been removed. Both species consume a similar diet of arthropods, but grasshopper mice also eat deer mice and other rodents.

Abundance of deer mice declined on both the control and treatment (removal) plots, but the decline was greater on the plots where grasshopper mice had not been removed. Deer mice increased their use

Figure 2.3 Red foxes are adaptable carnivores that live in a variety of environments. When introduced to Australia to control over-abundant, introduced rabbits that were devastating native flora, red foxes created another problem by preying on much of the native fauna. Photo by Stephen DeStefano.

of shrub cover on plots where grasshopper mice were abundant, but did not shift their use of microhabitat on sites lacking grasshopper mice. Grasshopper mice rarely used these shrubby microhabitats, which suggests that shrubs may have been important protective cover for deer mice. In addition, other rodent species (Ord's kangaroo rats (*Dipodomys ordii*), western harvest mice (*Reithrodontomys megalotis*)) responded to the removal of grasshopper mice by increasing their numbers or colonizing sites where they had not been present before the removal experiment. Stapp concluded that predatory or aggressive interference, rather than competition, was responsible for the changes he observed in abundance and microhabitat use by deer mice and that this was an important contributor to the structure of the small mammal community in shortgrass prairie.

Additive experiments

Additive experiments are used to determine whether additions in one species will cause changes in the distribution or abundance of other species. Thus, they evaluate a different phenomenon than the one assessed with removal experiments: additive experiments determine whether increasing the abundance of neighbors *above present levels* will provide evidence of interference (if other species decline) or facilitation (if other species increase). Even if one of these patterns is evident, this only shows that interference or facilitation could potentially structure the community if neighbor densities increased to the level created in the experiment (Keddy 1989). If these densities are not observed in nature, conclusions should be drawn with considerable caution.

Additive experiments usually examine the effect of extant organisms on introduced individuals or populations, but not the reciprocal effect. As such, they evaluate only one facet of the interaction. In addition, they do not determine which of the (usually) many extant species are responsible for the observed effects on the introduced species. In fact, two (or more) extant species may have opposite effects on the introduced species that reduce or eliminate the ability to perceive an effect. For example, the performance of a plant species may be enhanced by the direct effects of an overstory tree species; nonetheless, the beneficial effects of the overstory tree would be undetected if individuals of the introduced plant are killed by herbivores associated with the tree.

CASE STUDY: ADDITION OF *DIPSACUS* TO OLD FIELDS

Werner (1977) sowed seeds of a biennial plant (teasel, *Dipsacus sylvestris*) into eight abandoned agricultural fields with varying levels of plant cover. She subsequently monitored seeded and unseeded plots for 5 years. As the biennial plant increased in size and density, the growth and density of nearby herbaceous dicots were reduced. In contrast, nearby grasses did not respond to the introduction of the biennial. The biennial colonized all fields except the one with high cover of quackgrass (*Agropyron repens*), which indicates that quackgrass interfered with colonization by teasel. This study is unusual because the investigator evaluated the effect of an introduced species on extant vegetation as well as the effect of extant vegetation on the introduced species.

Laboratory experiments

Laboratory experiments illustrate the potential effects of interactions between species under specific sets of conditions which cannot be produced in the field. Ability to manipulate a wide range of variables is the primary advantage of working in the laboratory (e.g., controlled environment chambers, greenhouses). Laboratory experiments can facilitate precise identification of the mechanism underlying a particular interaction by allowing many variables to be held constant while one variable (e.g., species density) is varied. Disadvantages include the limited scope and extreme unrealism of these experiments; regardless of how convincingly an interaction is demonstrated in the laboratory, extrapolation to real-world ecosystems is limited. Thus, results from laboratory experiments may not provide evidence of the existence of an interaction in nature, much less indicate the strength of such an interaction.

CASE STUDY: LABORATORY EXPERIMENT – AGE EFFECTS
ON PLANT INTERACTIONS

Size of individuals has a clear impact on negative interactions between plants. However, the effects of plant age (at a given size) on interactions are less clear. Aspinall and Milthorpe (1959) addressed the influence of plant age on negative interactions between two annual plants that are capable of coexisting: a common crop plant (barley, *Hordeum vulgare*) and an associated weed (white persicaria, *Polygonum lapathifolium*). The two species were grown in pure and mixed populations of varying densities in greenhouse pots.

Growth of persicaria was reduced in the presence of barley, but growth of barley was unaffected by persicaria. This would seem to favor barley over persicaria over multiple generations, and lead to competitive exclusion rather than coexistence. However, differences in plant phenology favored coexistence. Specifically, persicaria accounted for only 8–10% of total biomass in mixed populations between 0 and 8 weeks, regardless of plant density. At 8 weeks of age, barley ceases vegetative growth and allocates resources to reproductive structures. This enables persicaria to increase in mixed populations, especially in low-density mixtures. In the field, coexistence presumably occurs as a result of differential plant phenology in combination with relatively low densities of barley in some locales.

Field experiments

Relationships between the abundance of neighbors and performance of target plant(s) are interpreted in the same way in field experiments as they are in laboratory experiments: an inverse relationship indicates interference, whereas a direct relationship indicates facilitation. Unlike laboratory studies, there is a reference point: the current performance of the individuals or populations of interest (Keddy 1989). Field experiments offer results that can often be confidently extrapolated to similar ecosystems. However, precise identification of the mechanism underlying a specific interaction may be hampered by confounding between the factor of interest

Figure 2.4 The suppression of natural wild fires has contributed to the spread of mesquite in the arid grasslands of the southwestern United States and northwestern Mexico. Photo by Stephen DeStefano.

(i.e., neighbor density) and other factors that are inadvertantly altered when neighbor density is manipulated (e.g., disturbance levels, nutrient pools and fluxes, organisms associated with manipulated species).

CASE STUDY: FIELD EXPERIMENT – PARTITIONING ABOVE- AND
BELOW-GROUND INTERFERENCE

Dramatic increases in the density of woody plants have occurred in savannas and grasslands throughout the world (McPherson 1997; Scholes and Archer 1997). Reduced interference from extant grasses as a result of livestock grazing has been implicated in these changes in physiognomy. Van Auken and Bush (1997) designed a field experiment to test the effects of a dominant native grass (sideoats grama, *Bouteloua curtipendula*) on seedling growth of a widely distributed woody plant (honey mesquite, *Prosopis glandulosa*) that has increased in abundance in central Texas (Figure 2.4).

Stands of grass were established in 1 m^2 plots and adjacent plots of the same size were hand weeded. Grass roots were allowed to grow throughout the site, so the hand-weeded plots represented "gaps" only above ground. Three years after the grasses were sown, mesquite seeds were sown into subplots which excluded grass roots at various depths. In addition, one-half of the plots were shaded to simulate interference for light. Thus, the experiment effectively partitioned interference into above- and below-ground components.

Mesquite seedlings were harvested at the end of one growing season, and biomass was determined. Mesquite growth was strongly affected by the depth to which grass roots were excluded. In addition, mesquite growth was reduced in grass-covered plots relative to gaps. However, above-ground interference had no measurable effect on growth of mesquite seedlings. These results indicate that: (1) interference from native grass stands can markedly reduce growth of mesquite, and (2) this interference occurs primarily below ground.

CASE STUDY: FIELD EXPERIMENT – THE 10-YEAR CYCLE

Authors have written about periodic fluctuations in animal densities since the early part of the 20th century. There has been much speculation regarding their cause; hypotheses have been based on overpopulation, random fluctuations, and meteorological causes (e.g., sunspots) (Keith 1963). Within the last two decades, observational studies have led researchers to believe that food abundance, nutritional quality of forage plants, and behavior and genetics may play important roles (Sinclair *et al.* 1982; Keith 1983; Ward and Krebs 1985; Krebs *et al.* 1995; O'Donoghue *et al.* 1998).

Population ecologists have long been aware of a phenomenon termed the "10-year cycle" (Keith 1963). The 10-year cycle involves periodic and dramatic fluctuations in populations of snowshoe hares (*Lepus americanus*) in the boreal forest of North America. Hare density peaks every 8–11 years, and the fluctuations are regular, of high amplitude, and are synchronous over a broad geographical region (Keith *et al.* 1984). These tremendous changes in hare density apparently involve interactions with several other wildlife species, such as ruffed grouse (*Bonasa umbellus*), the predators of hares and grouse – notably lynx (*Felis lynx*), coyotes (*Canis latrans*), and several species of raptors – and the plants that make up the hares' diet (Rusch *et al.* 1972; Brand *et al.* 1976; Todd *et al.* 1981; Boutin *et al.* 1995; O'Donoghue *et al.* 1997, 1998) (Figure 2.5).

Long-term studies or field experiments designed to test mechanisms underlying the 10-year cycle have been reported only relatively recently. Vaughan and Keith (1981) conducted such an experiment in eight enclosures (about 3–6 ha in size) of natural habitat to measure the demographic response of hares to food shortages during the winter. They used two levels of hare density, averaging about four and 13 hares/ha at the start of each experiment, and two levels of food availability (high and low). They found that food shortage greatly affected the reproductive characteristics of adult hares, including the onset and termination of breeding, pregnancy rates, and ovulation and implantation rates. These changes corresponded to shorter breeding seasons and a reduction in mean natality. Survival of juveniles was also markedly reduced, although survival of adults was unaffected. Vaughan and Keith concluded that the results of their field experiments were consistent with the view that cyclic declines in snowshoe hare populations are initiated by winter food shortage. This model of the 10-year cycle thus starts with an increasing density of snowshoe hares, triggered by a recovery in food resources. Several species of predators then show positive numerical and functional responses to the large numbers of hares, and their populations increase. Meanwhile, hare numbers have increased to the point at which intensive and widespread herbivory reduces plant-based sources of food, and the hare population declines rapidly. The decline in hares is followed closely by a decline in predator populations, marked by reduced reproductive output and lowered survivorship of the young of most predator species.

DETECTING INTERACTIONS

The study of community interactions is challenging for both plant and animal ecologists, but animal populations offer additional difficulties. Many species of animals are secretive, mobile, and long lived. Relative abundance can be quite variable over time and influenced by many factors, both biotic and abiotic. Marking individuals is rife with potential biases and accurate estimation of animal densities is problematic, even though methods are improving greatly (e.g., Buckland *et al.* 1993).

These logistical and methodological problems exist even in the study of small mammals (rodents), which have small home ranges and are abundant relative to other mammalian species (Dueser *et al.* 1989). Even some of the most frequently cited studies on competition in small mammal communities may have serious statistical flaws, which led Dueser *et al.* (1989:111) to conclude that we are less certain about the role of competition in rodent communities than is generally believed.

One of the primary problems with past studies is the lack of replication (Box 2.2). Dueser *et al.* (1989) reviewed 25 North American field experiments of competition that included both treatment and control plots. They reported that interspecific competition was prevalent, but that detection of competition was affected by experimental protocols. For example, competition was more evident between enclosed populations than between populations on open grids and, more disturbingly, competition was more evident in unreplicated than replicated studies. The evidence on competition in rodent communities is thus substantial, but ambiguous, which makes the generality of this evidence "unknowable but suspect" (Dueser *et al.* 1989:117). Dueser *et al.* (1989:123) emphasized

(a)

(b)

Figure 2.5 Interactions between (a) snowshoe hare and (b) lynx (here shown fitted with a radio telemetry collar and ready for release) as herbivore and predator represent a classic example of wildlife population

Box 2.2 Replication

Replication is the application of a treatment, or set of treatments, to more than one experimental unit. Replication demonstrates that observed trends are consistent, thereby reducing the possibility that a trend has occurred by chance. It increases the precision of estimates and it provides an estimate of experimental error, which is needed for appropriate statistical analyses.

In many cases, replication is precluded by ethical or logistical concerns. For example, it may be infeasible to conduct a replicated experiment on the behavior of large carnivores in a field setting, or on the effects of acid rain on trophic dynamics in large lakes. In such cases, inferences are restricted to the observed populations and considerable caution should be exercised when interpreting statements about broader populations (in these cases, carnivores or lakes not included in the study). The application of inferential statistics to data resulting from these studies is forced to rely on inappropriate error terms (i.e., sampling error is assumed to represent experimental error). As a result, these studies rely on pseudoreplication (Hurlbert 1984). The reader is encouraged to understand the causes and consequences of pseudoreplication with a thorough review of Hurlbert (1984) and Ramsey and Schafer (1997).

the need for increased statistical power in studies of community interactions by: (1) increasing replication; (2) reducing within-treatment variation; (3) using appropriate test statistics, such as one-tailed tests and repeated-measures analysis of variance; and (4) testing for large effect sizes by selecting an abundant species or by implementing the simultaneous removal and addition treatments that maximize between-treatment differences in density.

MANAGING INTERACTIONS

Management objectives frequently include unnatural conditions or situations. "Natural" levels of production (e.g., of red meat or wood) often do not meet societal demands. Similarly, "natural" levels of resource

Figure 2.5 (*cont.*) dynamics. Hare and lynx populations, interacting with their environment and each other, show cyclic highs and lows about every 10 years. Photos by Stephen DeStefano.

quality often are deemed insufficient (e.g., size of antlers or individual trees). Thus, managers focus on the creation of "unnatural" conditions to favor greater quantity or quality of various resources derived from ecosystems.

Many of the techniques designed to increase the quantity or quality of resources for human use rely on alterations in environmental conditions as a mechanism to alter species interactions and, therefore, change species composition. For example, removal of forest canopies generates large changes in several environmental parameters: at the soil surface, light and wind speed increase and diurnal temperature fluctuations become more pronounced. These and other changes in the physical environment tend to favor recruitment of species that are short lived, fast growing, and high in photosynthetic tissue at the expense of species that are long lived, slow growing, and high in structural tissue. The resulting suite of "early successional" species provides excellent habitat for some game species (e.g., white-tailed deer) compared with dense, closed-canopied forests. Conversely, many plant and animal species are found primarily in old forest stands with closed canopies, and the distribution and abundance of these species will likely be affected negatively by these management approaches (Figure 2.6).

Interference

Interference is a critically important interaction on most sites, and, in many cases, interference between plants can be manipulated to achieve management objectives. Effective manipulation of interference to meet objectives requires knowledge of two factors: the limiting resource(s) and the influence of the environment on the interaction. Considerable volumes of literature identify resource limitations and illustrate plant–environment relationships at an appropriate depth for the management of some systems. In other systems, relevant experiments should be conducted in order to identify and elucidate these relationships.

In most arid and semi-arid regions, water is the most limiting resource during much of the year, which implies that manipulation of other resources will have little or no impact on interference (and, hence, community structure). In contrast, light commonly constrains plant growth beneath forest canopies; thus, increasing the amount of light reaching understory plants is the most effective strategy for increasing the productivity of understory plants. These examples are simple and

Figure 2.6 The structural complexity of late successional, or old growth, forest provides ecological niches for a wide array of plants and animals. Photo by Stephen DeStefano.

obvious, yet they are representative of many sites. On other sites, plant productivity and diversity are constrained by one or more additional factors. Identifying the factors that impose the greatest constraints on the growth and survival of specific plants is a necessary first step toward manipulating the outcome of interference.

The ability of plants to interfere with neighbors is dependent on the environment (Keddy 1989; Grace and Tilman 1990). For example, the ability of many plants with the C_3 photosynthetic pathway to interfere with the growth of C_4 plants is enhanced under conditions of cool temperatures and low light (Williams *et al.* 1999). Similarly, recruitment of woody plants within a stand of grass is enhanced by a series of years with above-average precipitation or by defoliation of grasses by herbivores (McPherson 1997; Scholes and Archer 1997). Recognition of environmental effects on competitive interactions may allow manipulation of the environment to favor one plant or group of plants over others.

Herbivory

Herbivory represents an interaction and, in some cases, a management technique. Native herbivores associated with all wildland ecosystems remove an enormous amount of biomass each year and may constrain the establishment and growth of some species (Evans and Seastedt 1995). Thus, protection from these herbivores may be required to sustain populations of some plants. In effect, protection of desired plants from herbivores indicates that the primary constraint on plant establishment has been identified; protection represents an attempt to overcome this constraint. This strategy is particularly effective if plants require protection from herbivores only until they reach some minimum size or age. If native herbivores have been reduced via anthropogenic activities, reintroduction may represent a minimal-risk, low-cost strategy for reducing their forage plants (e.g., prairie dogs and woody plants; Weltzin *et al.* 1997; McPherson and Weltzin 2000).

Nonnative herbivores, notably livestock, have been introduced into most terrestrial ecosystems. The primary goals of livestock introduction are to convert biomass to a form suitable for human consumption and to produce other animal products (e.g., leather). Considerable volumes of literature address the consequences of livestock grazing and provide guidelines for maintaining the sustainability of this activity (e.g., Vallentine 1990; Heitschmidt and Stuth 1991; McPherson and Weltzin 2000). In general, livestock grazing requires minimal cultural input and represents a sustainable use of many areas if soils are inherently capable of recovering from livestock-induced compaction and if removal of nutrients does not exceed inputs from natural sources (e.g., precipitation, nitrogen-fixing organisms) (Figure 2.7).

Livestock can be manipulated to influence plant–plant interactions. Most large ungulates preferentially defoliate herbaceous plants, thereby enhancing the establishment and growth of woody plants. In contrast, goats tend to select woody plants and, hence, facilitate the spread of herbs. Managers can exploit these relationships to encourage desirable life forms and discourage undesirable ones. Historically, stocking rates of livestock on most natural areas greatly exceeded "sustainable" limits and thus impacted many populations of native flora and fauna. If applied judiciously, however, livestock grazing can be a useful method of manipulating vegetation in some areas (Box 2.3).

Figure 2.7 Livestock as well as wildlife are a potentially important ecological force that influences patterns of vegetation. Photo by Stephen DeStefano.

Seed dispersal

Several management activities represent attempts to overcome seed limitations on a site. As with herbivory, any attempt to increase the seed supply implies that seed dispersal is a primary constraint on establishment of desirable species. Constraints on seed availability can be overcome with a variety of techniques, ranging from direct sowing or planting to the attraction of animals likely to deposit the desired seeds.

Direct sowing is used by landscape architects, silviculturists, and revegetation specialists. Seeds of desirable species are frequently sown into areas after plants are removed by a disturbance (e.g., fire, overstory removal, road construction). The objective in these cases is to enhance the recruitment of desirable species by giving them a "head start" on undesirable species.

A less direct means of dispersing seeds involves feeding the seeds to domestic herbivores which are then released into target areas. Obviously, seeds must be resistant to deterioration or digestion by the animal. Seeds are defecated in feces, which provide a nutrient-rich, interference-free environment for germination and early establishment. This technique has been used with livestock to restore tropical dry forest in Guanacaste National Park, Costa Rica (Janzen 1986).

> **Box 2.3** Managing biological invasions
>
> The historical consequences of livestock grazing continue to influ-
> ence contemporary management decisions. Many vegetation man-
> agers refuse to consider livestock an appropriate tool for vegetation
> management because of historical "transgressions." For example,
> livestock are excluded from most national parks in the western
> United States: livestock are not native to these systems, and their
> presence appears to be inconsistent with the restoration and main-
> tenance of preColumbian plant communities.
>
> Grazing by livestock has contributed to extensive and wide-
> spread soil erosion and undesirable vegetation change in many of
> the world's ecosystems. Nonnative grazing animals caused, and
> continue to cause, reduced biological diversity. The livestock indus-
> try causes much economic and ecological harm with few societal
> benefits, and it appears to survive in many areas primarily because
> it offers a unique and colorful livelihood. However, livestock graz-
> ing is a useful tool for some site- and objective-specific goals of veg-
> etation management, including the maintenance of biological
> diversity.
>
> In the southwestern United States, nonnative annual grasses
> carpet the Sonoran Desert during years with above-average winter
> precipitation (Abbott and McPherson 1999). The associated in-
> crease in fine fuel enhances fire occurrence and spread, and most
> native plants are poorly adapted to fire. Long-lived succulents
> such as the giant saguaro cactus are particularly vulnerable.
> Opportunistic, short-term grazing by livestock may reduce the
> fine fuel load and prevent fires in these systems. Impacts of
> livestock grazing on this ecosystem presumably are minor relative
> to the long-term detrimental effects of fire (Abbott and
> McPherson 1999).

Finally, a very indirect means of overcoming constraints on seed
dispersal relies on passive assistance from native animals. Again, seeds
are resistant to deterioration or digestion, and they are defecated in desirable locations. Specifically, birds are attracted with artificial perches
that serve as recruitment foci for woody plants (McClanahan and Wolfe
1993; Robinson and Handel 1993). The effectiveness of perches is enhanced by providing seeds of desirable plants in bird feeders. As with the
application of much other ecological research, this strategy is not always

successful; perching structures typically increase seed dispersal, but may not overcome other constraints on woody plant recruitment (Holl 1998).

SUMMARY

It is difficult to conduct rigorous research on interactions that is relevant to management, particularly in animal populations. However, understanding interactions between organisms provides a foundation for studying collections of organisms, or communities. Thus, the scientific study of interactions underlies the quest to understand the coexistence of multiple species and, therefore, patterns of community structure. In addition, knowledge of interactions may facilitate the prediction of community response to changes in environment or land use. These topics are discussed in the following chapters.

3

Community structure

Questions about the number and diversity of species have entertained ecologists for generations. Many of these questions are pursued because they are academically intriguing, not because they offer significant insight into the management of ecosystems. For example, asking why so many species can coexist – a question deemed the Holy Grail of community ecology after Hutchinson (1959, 1961) posed it (Grace 1995) – is equivalent to asking why there are so many colors among birds. Species, like coloration, evolved as a result of selection pressures (including, for example, interference and facilitation). The fact that there are so many species is an inevitable outcome of the large number of generations and selective pressures evident since life evolved. Rather than pursuing questions with little applicability to management, this chapter will focus on the techniques used to describe community structure. Communities are a primary unit of management; therefore, understanding and describing community structure are requisite steps toward effective management.

THE COMMUNITY CONCEPT

Debate about the community concept began before "ecology" was formally defined in 1894 (Madison Botanical Congress 1894, cited in Langenheim 1995). The earliest discussions about collections of organisms included consideration of ecological interactions, with some scientists believing that interdependence among species was fundamental to the definition of communities (e.g., Mobius 1877) and others believing that no interactions were necessary to describe and discuss communities (e.g., Grisebach 1838). The debate was polarized by two North American plant ecologists: Frederic Clements, who promoted the idea that plant communities were superorganisms comprised of interdependent species (Clements 1916), and Henry Gleason (1917, 1926), a vocal proponent of Forrest Shreve's

Figure 3.1 Sonoran Desert in the foreground, with tidal and marine communities of the Gulf of California in the background, Sonora, Mexico. Photo by Stephen DeStefano.

(1915, 1922) idea that species are distributed in an individualistic manner along environmental gradients.

Modern discussions of the community concept generally focus on community structure (but see Drake 1990 for a functional view). Most contemporary definitions agree that species interact, although the degree of interaction remains the focus of some debate. Modern discussions of the community concept also tend to agree that communities are delimitable in space and time, that they are inseparable from climate, and that they are characterized by structural homogeneity. For the purposes of this book, we adopt these elements of the community concept (Figure 3.1).

DESCRIBING COMMUNITIES

Vegetation management is occasionally aimed at species populations (e.g., rare species), but it usually focuses on plant communities as the fundamental unit of management. Therefore, effective vegetation management depends on the ability to describe community structure objectively. Qualitative, coarse-level descriptions may be adequate for communication about, and management of, plant communities. These descriptions may simply name the dominant life form or species in the assemblage (e.g., semi-desert mixed shrub or Rocky Mountain mixed-conifer communities). Frequently, however, additional information is required to assess specific objectives. For example, a quantitative description of vegetation may be required to compare the efficacy of various management strategies in achieving desired states of vegetation. Quantitative descriptions may also serve as reference points for management.

The most inclusive interpretation of the term "community" would incorporate all coexisting organisms. Thus, plants, vertebrates, invertebrates, fungi, microbes, and other organisms that occupy a specific site comprise the ecological community. As this book builds on a foundation of plant ecology and vegetation management, the emphasis in this chapter is on vegetation. Animals can exert dominant effects on plant community structure, just as vegetation dominates the structure and composition of animal communities. Examples of the former include snowshoe hares in the boreal forest (as discussed in Chapter 2), ungulates in some grasslands of Africa and North America, white-tailed deer (*Odocoileus virginianus*) in urban settings in the northeastern United States, and livestock throughout much of the arid western United States. Other examples include the distribution of seeds by birds and sciurids and soil turnover by fossorial mammals and invertebrates (Figure 3.2). Concepts and techniques discussed in this chapter have been appropriately applied to collections of plants and assemblages of animals; they are suitable to various other applications, at fine or coarse taxonomic precision.

General goals of community ecology include: (1) description, as a basis for comparison with other sites or the same site at other times; (2) identification and elucidation of relationships between sites, species, and/or environmental variables; (3) identification of members of discrete classes; and (4) prediction of species composition. Most of these goals are addressed with pattern-oriented models. For example, typical tools used to address the first goal include standard community-based descriptors (e.g., indices of diversity, models of community organization). Ordination may also be used to accomplish the first goal, and it is the customary analytical approach used to address the second goal; additionally, ordination is useful for stimulating hypotheses about community structure. Classification is the typical approach used to address the third goal, which is frequently associated with mapping exercises or evaluation of the preservation status of plant communities. Process-oriented models are the primary formal tools used to address the fourth goal, although trial and error is the approach most commonly used by managers.

Figure 3.2 Eastern gray squirrels, as well as many other species of mammals and birds, can distribute the seeds of plants via food caching or hoarding. Photo by Stephen DeStefano.

Data collection

As already mentioned, plant communities can be described adequately in the absence of quantitative data. Thus, descriptors such as "sagebrush-grass" and "ponderosa pine/bunchgrass" may be adequate for many purposes. These coarse-scale descriptors are also often adequate to communicate the assemblage of vertebrate wildlife that one could expect to find: pocket gophers, sage sparrows (*Amphispiza belli*), black-tailed jackrabbits (*Lepus californicus*), coyotes, and pronghorn (*Antilocapra americana*) are found in "sagebrush-grass" communities, whereas western gray squirrels (*Sciurus griseus*), western tanagers (*Piranga ludoviciana*), northern goshawks, and mule deer (*Odocoileus hemionus*) are among the members of the "ponderosa pine/bunchgrass" community. In some cases, management is facilitated when communities are described in greater detail.

General considerations

Many texts have been developed specifically to facilitate the collection of data in an objective and repeatable manner (e.g., Mueller-Dombois and Ellenberg 1974; Grieg-Smith 1983; Kershaw and Looney 1985; Causton 1988; Ludwig and Reynolds 1988; Bonham 1989; Kent and Coker 1992). These texts indicate that data collected to describe plant communities should adhere to certain standards with respect to appropriateness, homogeneity, objectivity, and efficiency. Each of these elements is discussed below.

Data should be appropriate with respect to the character of the community, the investigator's research purposes, and plans for subsequent data analyses. For example, estimating the tree density in eastern deciduous forest requires a quadrat size several orders of magnitude larger than that required to estimate plant density in Californian annual grassland, simply because of differences in plant size. Similarly, if analysis of community structure is targeted at hypothesis generation, then overly quantitative assessments cannot provide the necessary broad coverage and may actually stifle the generation of hypotheses. Different objectives also dictate different intensities of data collection: objectives may include, for example, causal analysis of vegetation structure or feasibility of a site for a specific land use (e.g., recreation), with concomitant differences in data collection. Finally, data collection should not occupy a disproportionate amount of time and resources beyond the specific objectives of the research.

Samples should be homogeneous in structure and composition to ensure that each quadrat does not represent a different community. Because plant communities are not naturally delimited, subjective decisions must be made about which areas to include. Before sampling ponderosa pine/bunchgrass communities, the investigator must determine which parts of the landscape fall into this community. Sites vary in degree of homogeneity and in the type and severity of disturbances (Figure 3.3); historically, ecologists have selected the most uniform and least disturbed areas for subsequent study, although this may be changing as scientists recognize the importance of conducting research in areas that are more broadly representative of the planet's ecosystems. Also, scale is potentially problematic, since plants are rarely distributed randomly – they are patterned on several scales, so that a particular quadrat size and sample size will be adequate for some species and environmental factors, but too small to be representative for others, and too large to be homogeneous for others.

Animal ecologists must also consider the mobility and dispersal capabilities of animals in their studies of communities. Some animals are restricted in distribution; consider, for example, northern spotted owls (*Strix occidentalis*), which inhabit mature coniferous forests from central and coastal California, through western Oregon and Washington, and into southwestern British Columbia. In contrast, the great horned owl (*Bubo virginianus*) is found throughout North America, in all types of vegetation (Figure 3.4). Even the relatively restricted range of the northern

Figure 3.3 Gaps in the forest canopy caused by processes such as windthrow are common in most forests. These kinds of natural perturbations should be represented in the description of plant communities. Photo by Stephen DeStefano.

spotted owl encompasses several plant communities, such as the redwood forests of California, Douglas-fir/western hemlock forests of western Oregon, and mixed coniferous forests of central Washington. Animal ecologists are thus faced with additional challenges in defining communities; for example, Douglas-fir/western hemlock forests do not incorporate all of the forest types inhabited by northern spotted owls. In addition, deciding the scale at which to conduct research at the community level is very important; several authors have articulated concerns about scale – specifically the size of study areas (Smallwood 1995; Blackburn and Gaston 1996; Smallwood and Schonewald 1996).

There are many different sampling procedures, so selection of any one is subjective. However, once selected, sampling methods should be objective, standardized, and repeatable by other researchers. Methods should be described in an unambiguous and operational manner to facilitate comparisons across treatments, years, and data sets.

Finally, data collection should be an efficient process, to maximize the amount of information gained per unit of time, effort, and resources. This criterion conflicts with the previous three; in fact, most criteria will conflict with each other in many studies.

Despite the inability to devise a perfect protocol for any ecological study, the ultimate selection of sampling procedures should consider at

Figure 3.4 Great horned owls are widely distributed across North America and breed in a variety of environments, from boreal forests to deserts. Photo by Stephen DeStefano.

least the following components: objectives, scope, required accuracy of the study, kinds of communities sampled, type of environmental and historical data needed to complement vegetation or animal data, requirements to allow comparison with other studies, requirements for valid application of anticipated data analyses, and practical limitations. In addition, personal preferences of the investigator will influence the sampling protocol.

Typical sampling methods

Communities are usually comprised of several taxa, and one goal of community sampling is to describe the abundances of various taxa. Description is generally attempted at the species level of resolution. Several attributes can be used to signify abundance of a species, and the results of community analyses depend on the attribute selected (Smartt *et al.* 1974; Podani 1984; Kenkel *et al.* 1989; West and Reese 1991; Mehlert and McPherson 1996; Guo and Rundel 1997). Practical limitations (e.g., time, money) often constrain the number of attributes that can be sampled, which ensures that selection of an appropriate attribute is an important issue. Typical attributes used to describe the abundance of plant species include density, biomass, cover, and frequency. Typical attributes used to describe abundance of animal species include density and frequency.

Density (number of individuals per unit area) can be used with individuals that are easily distinguished. This precludes density as a basis for evaluating the abundance of many clonal organisms (e.g., sod-forming grasses or woody plants that resprout when burned or cut). The use of density assumes that all individuals are equally important, unless taxa are subdivided on the basis of size (*sensu* Prodgers 1984). Density estimation based on the point-quarter method (Cottam and Curtis 1956) is often used for plants, particularly large woody species. Density estimation based on distance sampling techniques (Buckland *et al.* 1993) may be the most effective and efficient method for determining the density of many vertebrate populations, especially species that are distributed sparsely across large geographical areas. Point counts for birds (Ralph and Scott 1981; Bibby *et al.* 1992) and capture–recapture techniques for small mammals (Pollock *et al.* 1990) are also used frequently (Box 3.1).

Biomass (mass per unit area) can be determined directly for some growth forms (e.g., herbs, small shrubs) and can be estimated for others. Biomass sampling is usually restricted to the above-ground portion of plants. Biomass is the most time-consuming and labor-intensive attribute to evaluate. However, biomass accounts for differences in size and growth form between taxa or individuals, and is therefore preferred by many community ecologists.

Cover (percentage of ground covered) can be estimated and expressed in terms of foliage or plant bases. Foliar cover indicates the total amount of light a plant is capable of intercepting, and is often applied to plants that are smaller than observers. Basal cover indicates the percentage of soil occupied by a plant, and is often applied to trees (e.g., diameter at breast height, commonly used by foresters). Basal cover is also used with herbs. Cover may be estimated along a line (line-intercept method, Canfield 1941), at a series of points (point sampling, Levy and Madden 1933, Goodall 1952, 1953), or via visual estimation within quadrats. In the latter instance, cover is often estimated within classes (Daubenmire 1968). Various scales are used for cover classes, with the octave scale

Box 3.1 Capture–recapture models

Capture–recapture (CR) methodology has a long history in ecology (Seber 1982; Pollock *et al.* 1990). In about 1930, Petersen and Lincoln independently used a simple form of CR to estimate the abundance of fish and waterfowl population size, respectively. CR has since been modified to estimate survival rate, which is the proportion of animals that remain alive after a time period or interval has passed. In its simplest form, CR is based on a proportion of marked individuals alive at time t that are still alive at time $t+1$. Abundance can be estimated with two capture occasions: a sample of x animals are captured and marked at time t, followed by another sample of animals captured at time $t+1$. The latter group will consist of unmarked individuals and previously marked individuals. The "Lincoln–Peterson index" is based on the formula $N = CM/R$, where M is the number of animals marked during the first capture occasion, C is the number of both marked and unmarked animals captured during the second occasion, R is the number of marked animals captured during the second occasion, and N is population size.

The basis for CR estimates lies with marked individuals – for example, small mammals with ear tags, fish with fin tags, or geese with neck bands. These marks can be simple batch marks (i.e., all individuals captured in a particular capture period receive the same mark, such as a colored tag or paint spot, that identifies them as members of that cohort but not as individuals). More sophisticated modeling can be done, however, if all individuals are marked uniquely, such as with numbered tags or individually color-coded marks. Marked animals are subsequently recaptured, or reobserved if they have visual marks that can be detected without actual capture.

CR has been used more recently to estimate survival (how many animals survive during a time interval), rather than simply abundance. These are the Cormack–Jolly–Seber models, based on the work of these independent researchers. To estimate survival, a minimum of three capture occasions is required; for many species of vertebrates, annual survival is the parameter of interest, so a capture occasion is defined as 1 year, although other periods such as days, weeks, or months could be used. The most reliable information will be attained from studies that run much longer than the requisite

3 years, however. For example, Forsman *et al.* (1993) determined the survival of northern spotted owls from several studies of 4–10 years.

Multiyear studies involve repeated capture occasions. For example, when a study begins in year 1, a sample of animals is captured, marked, and released back into the environment. In year 2, another sample of animals is captured; the unmarked individuals are marked and released, and the previously marked individuals have their numbers recorded and are then re-released. This process is repeated each year; over a period of time, each individual animal will have its own capture history, which is a series of 0s (for not captured) and 1s (for captured). A capture history matrix represents individual animals (rows) and whether or not they were captured during each time interval (columns). A capture history matrix could thus be:

```
11000
11111
01011
01000
00101
```

where the first animal was captured on occasion 1, recaptured on occasion 2, and never seen again during the remaining three periods of the study.

Recapture probability (p) can be estimated from the capture history matrix; p is considered a "nuisance parameter" because it does not reveal anything biologically meaningful about the population, but it is necessary to estimate survival. Many assumptions are involved in the use of CR methodology (Seber 1982; Pollock *et al.* 1990). For example, some models assume that populations are closed (i.e., there are no births, deaths, immigration, or emigration); clearly, this is a difficult assumption to meet and more sophisticated models deal with open populations. As with any modeling effort, adherence to assumptions is critical.

being especially common (Table 3.1). The results of community analyses are dependent on the scale that is being used (e.g., Jongman *et al.* 1987:27–8).

Frequency (percentage of quadrats occupied) can be evaluated more quickly than other attributes. However, frequency is sensitive to the size of quadrat used, which substantially constrains comparability

Table 3.1 *Conversion scale from percentage cover to octave scale*

Cover (%)	Octave value
0	0
0.01–0.49	1
0.50–0.99	2
1.00–1.99	3
2.00–3.99	4
4.00–7.99	5
8.00–15.99	6
16.00–31.99	7
32.00–63.99	8
64.00–100	9

Source: Adapted from Gauch and Stone 1979.

between studies. For example, most plant species occur infrequently in 0.5 m^2 quadrats, more frequently in 1 m^2 quadrats, and even more frequently in 2 m^2 quadrats. Different investigators routinely use different quadrat sizes, depending on objectives, kinds of communities, personal preferences, practical limitations, and other factors. Thus, frequency is rarely comparable between studies, even for the same species. In contrast, plant density, biomass, and cover can be compared between disparate ecosystems. For frequency counts of animal species, study area size can greatly influence the results (Blackburn and Gaston 1996; Smallwood and Schonewald 1996). In addition, study area selection in most wildlife studies is based on logistics to a greater extent than on study design. Although logistics (e.g., access) necessarily plays a role in study area selection, wildlife researchers should pay greater heed to establishing study areas in some sort of random fashion (e.g., simple random, stratified random, cluster sampling) (Ramsey and Schafer 1997).

Most quantitative studies of plant communities rely on some form of area sampling, as opposed to line or point sampling. Two caveats are warranted with respect to area sampling. First, plant distributions and environmental factors involve many spatial scales, so a specific size of quadrat may be appropriate for some species, too large for others, and too small for others. Second, accuracy of all area-based attributes usually can be improved by increasing the size and number of quadrats, but increased sampling effort is required. Use of distance sampling, in which distances from line or point transects to plants or animals are recorded, is a relatively new development based on the line-transect method (Buckland *et al.* 1993). Distance sampling overcomes the problems associated with area-based sampling methods, and allows some plants or animals to go undetected, except for those located on the line or point (i.e., objects on the line or point must be detected with a probability of 1) (Box 3.2).

Area sampling requires the selection of a specific shape and size of quadrat. Within-sample homogeneity is minimized in quadrats with a high proportion of area to edge (in decreasing order: circle, square, rectangle). In addition, the number of decisions about plants "in" or "out" of quadrats (hence, this source of error) is minimized in these quadrats. Conversely, homogeneity between samples is minimized with long, narrow quadrats (Brown 1954; Gauch 1982; Pielou 1984).

Box 3.2 Distance modeling

One of the most fundamental questions faced by research ecologists and resource managers is also one of the most difficult to answer: "*how many are there?*" Andrewartha (1961) recognized the importance of "how many" when he refined a definition of ecology as *the scientific study of the distribution and abundance of organisms.* Krebs (1972) took this a bit further in defining ecology as *the scientific study of the interactions that determine the distribution and abundance of organisms.* Determining the abundance of rare or secretive animals is difficult, but even large, obvious animals such as elk or elephants present a challenge. Plants are also not immune; estimating abundance is difficult for most organisms in natural environments.

Distance sampling is a technique that has been developed to aid biologists in determining abundance (Buckland *et al.* 1993). Line or point transects are the primary distance methods, and the sample data consist of the set of *distances* from the line or point to the plant, animal, or object of interest. Distance sampling is an extension of plot sampling, in which it is assumed that *all* objects within sample plots are counted, which is a very difficult assumption to meet under most field conditions. Distance sampling allows the researcher to miss some objects, except those that are directly on the line or point (those objects must be detected with a probability of 1).

Distance sampling takes into account the fact that the size of the sample area is sometimes unknown, that many objects away from the transect line or point may not be detected (thus distances are *sampled*), and that there is a tendency for detectability to decrease with increasing distance from the line. This "*detection function*", or $g(y)$, is the probability of detecting an object, given that it is a distance y from the random line or point. It is written as

$$g(y) = \Pr(\text{detection} \mid \text{distance } y).$$

where y is the perpendicular distance from the line, or the radial distance from the point. Graphically, distance functions can vary, but can often appear as half of a bell-shaped curve, illustrating a high probability of detection close to the line, with decreasing probabilities of detection as one moves away from the line.

As with all methods, there are critical assumptions and details of technique that must be followed. Despite these caveats, distance sampling offers a reliable method for answering a difficult question: *how many are there?*

Determination of the appropriate quadrat size may be quantitatively evaluated with two distinct approaches, each of which relies on the collection of data prior to initiation of the study: species–area curves or statistical methods. Species–area curves (Arrhenius 1921) are generated from series of nested plots, with a goal of sampling the area at which few or no additional species are added within the same ecosystem. Various estimates have been developed to predict the actual number of species from observed data, and therefore determine the appropriate quadrat size for sampling. The questionable reliability of these estimates (Hayek and Buzas 1997) suggests that selection of quadrat size based on species–area curves is best done subjectively. The most common statistical method involves the calculation of means and variances of common species for various quadrat sizes. This method can also be used to determine the appropriate number of quadrats needed to achieve a specified level of precision. A common form is (Eckblad 1991):

$n = t^2 s^2 / (\bar{X} k)^2$,   where
$n =$ number of quadrats needed,
$t =$ value of Student's t,
$s^2 =$ variance of the sample,
$\bar{X} =$ sample mean,   and
$k =$ desired accuracy.

For example, if the investigator desires that the mean be estimated within 10% of the true value, then accuracy is set to 0.1. If prestudy sampling of 15 samples produces a sample mean of 9.5 and a sample variance of 14.2 and if the investigator specifies a level of accuracy of 0.1 (i.e., the investigator desires an estimate of the mean that is within 10% of the true value), then:

$$n = t^2 s^2 / (\bar{X} k)^2 = 2.145^2 (14.2) / [(0.1)(9.5)]^2 \cong 72.$$

Note that the value of t (2.145) assumes $p = 0.05$ with 14 degrees of freedom.

Despite the widespread recognition and ease of application of these quantitative approaches, most investigators rely on custom to determine quadrat shape and size. A primary advantage of this reliance on convention is a high level of comparability between studies. A primary disadvantage is that estimates within a study may be inaccurate.

Data management

Absolute abundance of species may be less important than relative species composition. Relative abundance may be calculated for each measure of abundance, and is commonly calculated for density, frequency, and cover:

$RA_i = AA_i / \sum AA_i$, where
RA_i = relative abundance of species i,
AA_i = absolute abundance of species i, and
$\sum AA_i$ = total absolute abundance of all species.

Species abundance data commonly possess several characteristics that may impede subsequent analyses. For example, variances usually increase concomitantly with means, which violates the assumption of homogeneous variances required to conduct analysis of variance (ANOVA). Also, the presence of one or a few uncommon species in a data set may significantly influence the results of multivariate analyses. Finally, the results of univariate and multivariate analyses are sensitive to the presence of influential observations (i.e., "outliers").

ANOVA need not be employed to describe community structure – hence, homogeneous variances are not necessarily problematic. However, ANOVA is useful for comparing abundance and diversity of species between sites. Such comparisons are ill-advised with data that violate assumptions of ANOVA. One potential solution to this problem is the use of transformations to produce data which meet the assumptions of ANOVA. Any of several transformations may be appropriate, depending on the nature of the data (e.g., Bartlett 1947; Ramsey and Schafer 1997).

Log transformations have been proposed to homogenize variances, particularly with data represented as proportions:

$$y_{transformed} = \log (y_{original}).$$

Because most ecological data sets contain several values of 0, a more commonly used form is:

$$y_{transformed} = \log (y_{original} + 1).$$

The log-odds transformation is becoming increasingly common because transformed values can be interpreted directly as probabilities (Menard 1995):

$$y_{transformed} = \log [y_{original}/(y_{original} + 1)].$$

Log transformations have been incorporated into some protocols for data collection (e.g., van der Maarel 1979).

Removal of uncommon species is often recommended before proceeding with analyses. Specifically, species that occur with $\leq 5\%$ frequency in the entire data set are often removed (Gauch 1982). Effects of such species removals have not been studied, which suggests the adoption of a conservative stance: analyses should be conducted with *and* without inclusion of uncommon species. Simply removing these species prior to analyses may obscure relevant patterns in the data. Similarly, the elimination of influential observations is recommended only when there are ecologically meaningful rationales for discarding them.

Quantifying diversity

Relatively few descriptors can be used to compare communities that have dissimilar species composition. One such descriptor is diversity. Within-community diversity (i.e., alpha diversity, or α) can be expressed in several different ways; a few of these are discussed in the following sections.

Richness

Richness (number of species, expressed as s) is the most simple and straightforward measure of diversity. However, there are conceptual and theoretical disadvantages associated with the use of s as a single indicator of diversity (Figure 3.5).

One potential problem associated with richness is that the value of s calculated for a particular community is dependent on the area sampled.

Figure 3.5 A tremendous diversity of life, which can be described with measures of species richness and evenness, exists in tropical regions, such as this forest in Belize. Photo by Stephen DeStefano.

Therefore, comparisons between communities require the same sampling area. Transformations have been proposed to minimize this problem and therefore facilitate comparisons between communities:

$d = s/[\log (\text{area sampled})]$ (Odum *et al.* 1960),   or

$d' = (s - 1)/[\log (\text{area sampled})]$ (Margalef 1951, cited in Peet 1974),
   where

d and d' represent the rate at which species are added with increasing area.

These formulae assume sampling based on area (i.e., quadrats). If data are collected on the basis of individuals instead of area, equivalent formulae are:

$d = s/[\log (N)]$,   or

$d' = (s - 1)/[\log (N)]$,   where

$N = $ sample size.

These forms are commonly used when conducting point samples of bird populations. Use of the denominators in these formulae assumes that there are logarithmic relationships between s and area sampled or N. This assumption may be inaccurate in some communities.

A second disadvantage of richness is its vulnerability to high sampling variability. This is particularly problematic with species that are not randomly distributed (e.g., rare species).

s can be affected by the relative distribution of species in the community, as illustrated by the following example. The expected value of s from a particular sample (Hurlbert 1971) is:

$$E(s) = \sum \left[1 - \binom{N - N_i}{n} \Big/ \binom{N}{n} \right], \text{ where}$$

N = total number of individuals in the community,
s = actual number of species,
N_i = number of individuals of the ith species,   and
n = sample size.

Given a population of $N = 990$ individuals, $s = 3$, perfectly even distribution of species (i.e., $N_1 = N_2 = N_3 = 330$), and a sample size of $n = 5$:

$$E(s) = 3 \left[1 - \binom{990 - 330}{5} \Big/ \binom{990}{5} \right]$$

$$= 3 \left[1 - \left(\frac{660!}{655! \, 5!} \right) \Big/ \left(\frac{990!}{985! \, 5!} \right) \right]$$

$$= 3 \, [1 - (660 \times 659 \times 658 \times 657 \times 656 / 990 \times 989 \times 988 \times 987 \times 986)]$$

$$= 2.6.$$

Sampling this community of 990 individuals at the specified sampling intensity indicates a richness of 2.6 different species. Given a second community of $N = 990$ individuals and $s = 10$, an uneven distribution of species (in this case, $N_1 = 900$ and $N_2 = N_3 = N_4 = \ldots = N_{10} = 10$), and a sample size of $n = 5$. This community is the same size as the previous one and the sampling intensity is identical. However, because of the uneven distribution of species in this community, the expected value of s is 1.4 different species. Thus, sampling the two communities indicates that the first community is more diverse than the second ($E(s)_I = 2.6$, $E(s)_{II} = 1.4$); yet, the actual richness of the second community is over three times that of the first community ($s_I = 3$, $s_{II} = 10$). In other words, sampling (which is necessary, because lack of time and money preclude censusing) may fail to produce an accurate representation of the number of species, and this limitation is particularly significant at low sampling intensities.

Finally, a simple example illustrates the most significant conceptual disadvantage associated with s. Consider two communities, each with five species (thus, $s = 5$ for each community). Community I has ten individuals of each species, and community II has 42 individuals of one species and two individuals of each of the other four species. Most people would consider community I to be more diverse than community II, although this is not reflected in s ($s_I = s_{II} = 5$). Thus, community I has a more equitable distribution of species: it has a higher evenness (equitability, heterogeneity).

These disadvantages illustrate that richness often does not adequately reflect diversity. As a result, several other indices of within-community diversity have been derived to capture and reflect differences in evenness. The two most common indices, Simpson's index (Simpson 1949) and the Shannon–Weaver index (Shannon and Weaver 1949), make the following assumptions: (1) all species in the community are represented in the sample (i.e., sampling intensity is adequate to "capture" all species in the community); (2) there is a population of size N from which we can draw an infinite number of samples without replacement (i.e., the population is infinitely large); and (3) these samples represent random samples of the population. These assumptions can be met reasonably well with appropriate attention to sampling methods.

Pielou (1966, 1975) criticized Simpson's index and the Shannon–Weaver index on theoretical grounds. Specifically, she concluded that the assumptions underlying these models could not be met with field data: all species in the community are virtually never represented in the sample, and, because plants are not distributed at random, quadrats do not represent random samples. Thus, the quadrat is not a sample taken from a population, but rather is a "community" in and of itself. These criticisms led Pielou and other ecologists to develop additional indices of diversity or evenness (Pielou 1966; McIntosh 1967, Hill 1973; Pielou 1975, 1977; Alatalo 1981; Molinari 1989; Nee et al. 1992; Camargo 1993; Bulla 1994; Hill 1997).

Despite the valid criticisms of Simpson's index and the Shannon–Weaver index and the derivation of numerous other indices, the former indices remain the most frequently used. Their widespread use, and the associated decline in derivation of new indices after the late 1970s, results from at least three factors: (1) all indices have disadvantages – Magurran (1988) provides an excellent summary of indices which illustrates this; (2) diversity and evenness are descriptors of community structure, but they do not necessarily provide information about the management or function of communities; and (3) these simple, early indices are adequate for the description and comparison of communities. Thus, familiarity and pragmatism are important factors in the selection and use of diversity indices. In short, the pursuit of diversity indices appears to be an intellectually bankrupt enterprise. Although the development of new indices continues among some ecologists, these indices are not being adopted and applied by field biologists.

Simpson's index

Simpson's (1949) index can be easily explained with a little reliance on probability theory. Suppose that it is possible to randomly select two individuals from a community of N individuals, s species, and N_i individuals in the ith species. Simpson's index is based on the probability that the first individual is the same species as the second individual. If the probability is high, then the community is not diverse. In particular, one formula for a random sample with replacement is:

$$c = [\Sigma N_i (N_i - 1)]/[N (N - 1)], \quad \text{where}$$
$$N = \text{total number of individuals in the sample,} \quad \text{and}$$
$$N_i = \text{number of individuals in the } i\text{th species.}$$

The parameter c represents the probability that two randomly selected individuals are the same species, and it is a statistically unbiased estimate. The analogous formula for sampling without replacement (i.e., individuals are harvested at the time of sampling, and are not placed back into the community) is:

$$\check{c} = \Sigma (N_i/N)^2.$$

In this formula, N_i/N represents a relative value of abundance which is commonly expressed as $\hat{\rho}_i$ (i.e., $\check{c} = \Sigma \hat{\rho}_i^2$). The parameter $\check{c}$ is the most common expression of Simpson's index, despite the fact that it is a statistically biased estimate.

Increased diversity in a sample is associated with a decreased probability that two randomly selected individuals will be the same species. Thus, to ensure that Simpson's index increases with increased diversity, $1 - c$ or $1/\check{c}$ are often used to express diversity. The latter form has an intuitive interpretation: it is the number of equally abundant species necessary to produce the same diversity as that observed in the sample. For example, consider the two simple communities discussed previously:

Community I has 50 individuals, evenly distributed among five species;

$$c_1 = 5 [10 (9)]/[50 (49)] = 0.18;$$
$$\check{c}_1 = 5 [(10/50)^2] = 0.2;$$
$$1/\check{c}_1 = 1/0.2 = 5.0.$$

The probability that two individuals selected at random from this community will be the same species is 0.18. In addition, the number of equally abundant species necessary to produce the same diversity as that observed in the sample is 5.0 (which is expected, given that the community is comprised of five equally abundant species).

> Community II has 50 individuals, with 42 individuals of one species and two individuals of the remaining four species.
>
> $c_{II} = [42\,(41)]/[50\,(49)] + 4\,[2\,(1)/50\,(49)] = 0.706$
>
> $\check{c}_{II} = (42/50)^2 + 4\,[(2/50)^2] = 0.712$
>
> $1/\check{c}_{II} = 1/0.712 = 1.4.$

The probability that two individuals selected at random from this community will be the same species is 0.706. The number of equally abundant species necessary to produce the same diversity as that observed in the sample is 1.4. In other words, community II is considerably less diverse than community I, even though they have an identical number of species (hence, identical values of s).

Simpson's index is more sensitive to the abundance of dominant species than rare species, and it has, therefore, been termed a "dominance" index. For example, one species ($N_i = 42$) in community II contributes 0.7056 to the total value of c_{II}, whereas the combined contribution of the other four species ($N_i = 2$) is only 0.0064: the most abundant species contributes over 99% to the total value of c_{II}. As a result, Simpson's index is relatively insensitive to sampling variability if the sample is adequate to represent dominant species.

Shannon–Weaver index

The Shannon–Weaver index (Shannon and Weaver 1949) is synonymous with Shannon's index and the Shannon–Wiener index. Like Simpson's index, the Shannon–Weaver index is based on probability theory. The most common form of the Shannon–Weaver index is:

$$H' = -\Sigma\,\hat{p}_i\,(\ln\,\hat{p}_i), \quad \text{where}$$
$$\hat{p}_i = N_i/N.$$

This form of the Shannon–Weaver index relies on proportions of species, as reflected in the use of $\hat{p}_i$. It is derived from the formula that expresses the probability of selecting all N_i individuals of all s species, P:

$P = \sum (N_i/N)^{Ni}$,   which can be expressed as

$P = (N_1/N)^{N1} \times (N_2/N)^{N2} \times \ldots \times (N_s/N)^{Ns}$,   where

N_i = number of individuals in the ith species,   and

N = total number of individuals in the sample.

H' is the most common form of the Shannon–Weaver index. The quantity $e^{H'}$ is analogous to $1/\check{c}$, a form of Simpson's index; it is the number of equally abundant species necessary to produce the same diversity as that observed in the sample. For example, consider the two simple communities discussed previously:

> Community I has 50 individuals, evenly distributed among five species;
>
> $H'_I = -5\,[(10/50)\ln(10/50)] = 1.6094$,
>
> $e^{H'}_I = e^{1.6094} = 5.0$.

The number of equally abundant species necessary to produce the same diversity as that observed in the sample is 5.0. This is to be expected, given that the community is comprised of five equally abundant species.

> Community II has 50 individuals, with 42 individuals of one species and two individuals of the remaining four species.
>
> $H'_{II} = -\{(42/50)\ln(42/50) + 4\,[(2/50)\ln(2/50)]\} = 0.6615$;
>
> $e^{H'}_{II} = e^{0.6615} = 1.94$.

The number of equally abundant species necessary to produce the same diversity as that observed in the sample is 1.9. In other words, community II is considerably less diverse than community I, even though they have an identical number of species (hence, identical values of s).

The Shannon–Weaver index increases logarithmically with increases in s. In theory, it is relatively sensitive to the abundance of rare species (Peet 1974). In practice, the Shannon–Weaver index is usually very highly correlated with Simpson's index, and frequently only one of the two indices is reported in research papers.

Models of community structure

Although plants comprise over 99% of the biomass of the Earth, there are no general predictive models of community structure or models for predicting the response of communities to changes in environment or land use (Prentice and van der Maarel 1987; Keddy and MacLellan 1990). The many models developed thus far have failed to generate a theoretical basis for community ecology, and they have been applied infrequently to the management of plant communities. Models of pattern predominate, and these models may be especially well adapted to management applications for both plants and animals.

Models of community organization

The number of species (i.e., richness) and the relative abundance of each species (i.e., evenness) are integral properties of communities. In addition, they are among the few attributes that can be used to compare disparate communities. These properties form the basis for the consideration of diversity relationships at a more general level than individual communities. Specifically, they lead to the idea that dominance/diversity relationships can be explained with a model of community organization. Several such models have been developed.

The niche pre-emption model (synonymous with geometric model: Motomura 1947; Whittaker 1972) assumes that the proportion of total available resources used by a species is reflected in the abundance of the species. The most abundant species captures some proportion of the resources, and the second-most abundant species uses a similar proportion of the remaining resources, and so on, for all species in the community. This model predicts a logarithmic relationship between species rank (i.e., most to least abundant) and species abundance, and is mathematically represented as a geometric series. It appears to describe dominance/diversity relations reasonably well in conifer forests with low richness or species-poor strata within communities (Whittaker 1965; Whittaker and Niering 1975), in agricultural fields within a few years after abandonment (Golley 1965; Bazzaz 1975), and in particularly arid or cold environments (Whittaker and Niering 1975; West and Reese 1991). In these situations, dominance tends to be strongly developed and species may interact directly and negatively.

The niche pre-emption model has been interpreted on the basis of environmental "harshness" (Whittaker 1965). Specifically, the presence of relatively few species implies the existence of many factors that act to limit productivity and diversity. As a result, species are forced to compete for scarce resources, and this competition underlies community structure.

In contrast to the niche pre-emption model, the general log-normal model (Preston 1948) implies that negative interactions have minimal impact on community structure. The log-normal model describes communities that are typified by many species of intermediate abundance; few species are either rare or very common. In this situation, the relative abundances of species are assumed to be approximately normally distributed, and are therefore said to be governed by relatively independent factors (Whittaker 1965). According to May (1975), "the lognormal distribution is associated with [results] of random variables, and factors that influence large and heterogeneous assemblies of species indeed tend to do so in this fashion ... If the environment is randomly fluctuating, or alternatively as soon as several factors become significant . . . we expect the statistical Law of Large Numbers to take over and produce the ubiquitous lognormal distribution." Colinvaux (1986:676) takes the interpretation a step further, stating that the "prevalence of log-normal distributions shows that the relative distribution of animals and plants very often is determined by random processes." Negative interactions (i.e., competition) appear to be less important in the log-normal model than in the niche preemption model. Nonetheless, the conclusion that random processes cause communities to be structured in a certain way exemplifies the inappropriate – but common – inference of cause-and-effect based on a pattern.

The broken-stick model (MacArthur 1957) indicates that niches of species are limited by competition at randomly located boundaries. Competition is explicitly incorporated into this model, which was developed to interpret patterns of territorial birds. The broken-stick model seems to be most applicable to communities that are comprised of relatively few species that are taxonomically similar (May 1975; Colinvaux 1986). The broken-stick model is usually viewed as being "extreme" with respect to the control of community structure by competition, and the assumption that competition controls community structure is unwarranted in the absence of conclusive supporting evidence. The broken-stick model is usually expressed mathematically:

$P_r = (N/s) \sum [1/(s - i + 1)]$,   where

s = number of species,

N = number of individuals,

i = species sequence from least to most important,   and

P_r = abundance of species n.

For example, given $N = 100$ individuals of $s = 3$ species, the abundance of the first three species is 11.1, 27.8, and 61.1: species $1 = (100/3)$ $[1/(3 - 1 + 1) = 11.1$; species $2 = (100/3)[1/(3 - 1 + 1) + 1/(3 - 2 + 1)] = 27.8$; and species $3 = (100/3)[1/(3 - 1 + 1) + 1/(3 - 2 + 1) + 1/(3 - 3 + 1)] = 61.1$. (Note that $\sum P_r = 100 = N$.)

Several other models of community organization have been developed within the last two decades. In addition, statistical approaches have been used to compare the validity of models on some sites (Wilson 1991;

Watkins and Wilson 1994; Wilson *et al.* 1996). These comparisons have failed to produce a consensus regarding an optimal approach (Wilson 1993; Watkins and Wilson 1994; Wilson *et al.* 1996). The slowed pursuit of an optimal model of community organization probably reflects the fact that: (1) all models are appropriate for some domains, but none have universal application; (2) these models describe community structure, but they do not provide information about the function or effective management of communities; and (3) existing models are adequate for the description and comparison of communities. Recently developed models have not been adopted and applied by field biologists, which suggests that future models will be met with indifference by potential model users.

Ordination

Ordination encompasses a family of multivariate techniques which are used to summarize complex data and relate species and community patterns to environmental variables. Multivariate analyses are fundamental to the objective description of communities. Because many species respond simultaneously to environmental factors, adequate description of community structure relies on multivariate approaches. In contrast, approaches that focus on the response of a single variable (e.g., biomass, richness, abundance of a specific species) – or several such variables – are necessary but insufficient for the effective management of most communities.

DIRECT GRADIENT ANALYSIS

Direct gradient analysis (DGA) is used to display the distribution of organisms along gradients of important environmental factors. DGA was devised by Ramensky (1930) and Gause (1930), but was not widely used by ecologists until the 1950s, when it was promoted by Whittaker as an effective tool to describe community structure (e.g., Whittaker 1956, 1960, 1962; Whittaker and Niering 1965; Whittaker 1967).

In DGA, data are plotted directly against environmental axes (hence, direct gradient analysis). With respect to environmental influences on plants, these axes may be direct (e.g., temperature), indirect (e.g., elevation), or synthetic (e.g., drainage class). Similar axes could be applied to animal assemblages. Species, communities, and community-level attributes may be plotted along one to several dimensions. Thus, attributes are plotted along environmental axes, and the investigator selects the attributes to be displayed (dependent variables) and the environmental axes (independent variables) (Figure 3.6).

Some form of data smoothing is usually employed before the presentation of DGA results. The resulting curve is less "noisy" than the

Figure 3.6 Environmental gradients can be subtle, such as changes in soil nutrients, or relatively dramatic, such as this change in elevation from the Great Basin Desert to mountaintops in southeastern Oregon. Photo by Stephen DeStefano.

original data. A common approach is to weight the current datum twice as heavily as the previous and next observations:

$$\text{datum}_{\text{smoothed}} = [\text{previous datum} + 2(\text{current datum}) + \text{next datum}]/4.$$

The large body of DGA-based research conducted by Whittaker enabled him to draw the following conclusions about species distributions (Whittaker 1965, 1972, 1975):

1. The general form for the distribution of a species population along an environmental gradient is the bell-shaped curve. The center (mode) of a species population along such a gradient is not at its physiological optimum (i.e., fundamental niche) but is a center of maximum population success in competition with other species populations (i.e., realized niche). In most cases, the centers of species populations are scattered along the gradient in an apparently random manner.

2. Species do not form well-defined groups of associates with similar distributions, but rather are distributed according to the principle of species individuality. Each species is distributed in its own manner, according to its own genetic,

physiological, and population response to environmental factors that affect it, including the effects of other species.

3.   Along an environmental gradient, species populations form a population continuum or compositional gradient. This suggests that, in the absence of environmental discontinuities or disturbances, communities intergrade or are discontinuous with one another.

These conclusions led Whittaker to reject the "community-unit" (Clementsian) hypothesis of community organization.

Whittaker's conclusions were strongly influenced by his assumption that species distributions are typified by bell-shaped curves. Whittaker routinely smoothed the curves of species distributions by hand, which facilitated widespread acceptance of this assumption among Whittaker's contemporaries. The assumption that most species distributions are bell shaped was challenged shortly before Whittaker's death in 1980 (Austin 1976). It is now generally accepted that bell-shaped curves of species distributions are less common than other curve shapes (Werger *et al.* 1983; Austin and Gaywood 1994).

DGA is firmly grounded in classical plant ecology (cf. Jack Major's (1951) functional factorial approach: vegetation is dependent on topography, organisms, time, soil, and climate). It is useful for summarizing and presenting data, particularly when important environmental variables are readily appreciated and measured and when objectives include direct, integrated use of environmental data. DGA is also valuable for the generation of hypotheses. However, the use of data smoothing may be misleading, especially when done by hand. In addition, DGA is highly subjective and it is inherently circular: a relationship between species distributions and the environment is observed and quantified, which leads inevitably to the conclusion that there is a relationship between the community structure and the measured environmental variables. Furthermore, the DGA-based conclusion of a vegetation continuum results directly from the arbitrary, subjective sampling employed in this approach.

INDIRECT ORDINATION

Indirect ordination (hereafter, ordination) was developed to overcome the subjectivity and bias associated with DGA. It is particularly useful if relationships between environmental variables and community structure are not obvious.

In ordination, axes that represent the major directions of environmental and community variation are sought from computations on the data; that is, data are summarized and patterns are sought using only species-abundance data (or sometimes species-occurrence data). Environmental interpretation is usually a subsequent and independent step in the analysis. Specific objectives of ordination are: (1) to summarize community data by producing a low-dimension ordination diagram (typically 1–4 dimensions) in which similar species and samples are close together and dissimilar entities are far apart; and (2) to relate species and community patterns to environmental variables. These two objectives reflect the two common philosophies about ordination: (1) it is a technique for matrix approximation, and (2) there is an underlying (latent) structure in the data that is dictated by the environment. Adoption of the former philosophy implies that multispecies assemblages are too complex to comprehend, and ordination is a tool for approximating nature's complexity. Adherence to the latter philosophy assumes that occurrences and abundances of all species are determined by a few unknown environmental variables according to a simple response model. In either case, ordination is used to display patterns of community structure and it can be used to generate hypotheses about the factors underlying that structure.

Several specific algorithms have been developed to conduct ordination. Interpretation of the results is broadly similar between these techniques. Therefore, this section will focus on the interpretation of ordination results and assumptions shared by a majority of the commonly used techniques. A geometric explanation is provided for a widely used ordination algorithm – principal components analysis (PCA). Ordination techniques employ eigenanalysis (a matrix algebra technique) instead of a geometric approach; however, an explanation of ordination based on geometry is more intuitive and approachable than one based on eigenanalysis, and the resulting level of understanding is sufficient for most ecological applications. More formal mathematical treatments of PCA and other algorithms, including discussions of eigenanalysis, are provided by several authors (e.g., Pielou 1977; Orlóci 1978; Gauch 1982; Manly 1986; Jongman *et al.* 1987).

Interpretation of ordination results is relatively straightforward. Values derived for species (i.e., species scores) and/or quadrats (i.e., quadrat scores, sometimes termed sample scores or stand scores) are typically plotted in two or three dimensions. Similar species and similar quadrats occur near each other, whereas dissimilar entities are far apart. The quadrat score is a simple, single-number representation of all the vegetation in a quadrat.

Distribution of species scores and quadrat scores consequently serves as the basis for interpreting environmental effects. Specifically, axes (i.e., dimensions) are interpreted as the factors underlying community structure. Knowledge of species' natural histories greatly enhances the interpretation of these axes. Ideally, few axes account for much of the variability in the data and these axes can be explained on the basis of ecological knowledge. However, there is no guarantee that few axes will shed appreciable light on community structure, because ordination is a mathematical exercise – it is not designed to evaluate ecological information beyond the bounds of a simple computational algorithm. Obviously, ordination results do reflect relationships between species and sites most of the time, as evidenced by the considerable application of ordination algorithms.

Ordination algorithms require axes to be mathematically uncorrelated with one another (i.e., orthogonal to each other), which minimizes the probability that they will be highly correlated from an ecological viewpoint. For example, if the first axis is interpreted as a gradient of litter depth, it is unlikely that the second axis will be interpreted as a gradient of organic carbon content, because litter depth and organic carbon content tend to be highly correlated.

In addition to the interpretation of axes based strictly on knowledge of species natural histories, a more formal numerical approach may be used. Specifically, correlation coefficients may be calculated between quadrat scores on a particular axis and the corresponding environmental variables. The strength of these associations may indicate the environmental variables underlying community structure. For example, a Pearson product-moment correlation coefficient of 0.9 between organic carbon content and quadrat scores on the first axis indicates that organic carbon is highly correlated with vegetation on the first axis, and suggests that organic carbon makes an important contribution to community structure. A subsequent field experiment may be used to test for a causal relationship between organic carbon content and community structure.

Environmental data can be included in the ordination, rather than treating environmental interpretation as a second step in the analysis. In this approach, values for environmental parameters are treated like values of species abundance, and the resulting "quadrat" scores associated with environmental variables are interpreted as if they are influencing community structure (Gauch and Stone 1979). However, because all variables included in an ordination influence the results, inclusion of environmental variables may reveal variability in the environment, thereby masking variation in community structure. Therefore, we suggest conducting an ordination based on species abundances, and then using correlation analysis to determine the strength of association between ordination axes and environmental variables.

Many ordination algorithms initially used in ecology assume that species are distributezd linearly along environmental gradients. Departures from linearity can produce a misleading view of the original data, and may interfere with appropriate interpretation. In particular, these departures from linearity are expressed by an "arch" in a two-dimensional ordination diagram with scale contraction at the ends of the gradient. In some cases, the "arch" effect and scale contraction are accompanied by involution, which incorrectly indicates similarity in dissimilar extremes along an environmental gradient (Figure 3.7).

The "arch" effect and associated problems are evident in analyses of simulated data sets, in which gradients are well-known and understood. They also appear to be present in many ordination diagrams derived from field data. Gradients that influence community structure in the field are usually unknown, or ordination would not be necessary. Therefore, it is difficult to differentiate between patterns in field data and artefacts that result from nonlinear data. Nonetheless, some ordination algorithms are more robust than others to nonlinear relationships between species abundances and environmental factors (i.e., they are not greatly affected by nonlinear relationships). For example, the "arch" effect produced by reciprocal averaging (RA; synonymous with correspondence analysis, CA) is less pronounced than the "arch" produced by PCA.

The widespread availability of inexpensive computers contributed to the development of several ordination techniques after 1970. Each new algorithm has been heralded as the best tool for quantifying community structure. Recently developed ordination algorithms assume that species have unimodal – rather than linear – response curves along environmental gradients. This idea echoes Whittaker's belief that "species are generally distributed with unimodal, often approximately Gaussian, distributions of abundance along underlying environmental gradients" (Peet *et al.* 1988:924). Although bell-shaped curves of species distributions are actually rare (Werger *et al.* 1983; Austin *et al.* 1994), they may be more common than the linear distributions assumed by PCA and RA. In addition, algorithms may be relatively insensitive to deviations from normality. Thus, recently developed ordination algorithms often produce ordination diagrams that are reasonably interpretable.

Detrended correspondence analysis (DCA) provides a means of eliminating the "arch" effect and the associated scale contraction (Hill and Gauch 1980) and is, therefore, a widely used ordination algorithm. However, the means by which DCA eliminates the "arch" effect and scale contraction is highly artificial (Pielou 1984; Minchin 1987), a factor which led Hill and Gauch (1980) to suggest that axes should be uncorrelated with the square and cube (and so on) of previously derived axes. Such "detrending-by-polynomials" was subsequently formalized by ter Braak and Prentice (1988) and incorporated within the computer program CANOCO (ter Braak 1988).

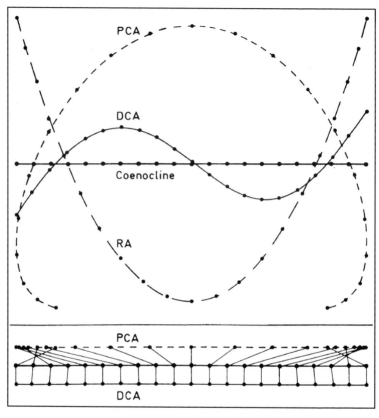

Figure 3.7 Representation of a simulated coenocline by three ordination techniques: centered principal components analysis (PCA), detrended correspondence analysis (DCA), and reciprocal averaging (RA). Upper panel shows two-dimensional solutions, with the first ordination axes on the horizontal axis scaled to the same width to facilitate comparison; second ordination axes are shown on the vertical axis and are scaled in proportion to their corresponding first axes. Axis polarity is arbitrary, but RA and PCA are presented with opposite polarities for clarity. Note that DCA ordination shows the least distortion of the original coenocline, RA distorts it into an arch, and PCA arches and involutes the original one-dimensional configuration. Lower panel shows that the first axis of PCA recovers the coenocline poorly, whereas the first axis of DCA is nearly a perfect match to the coenocline. The first axis of RA (not shown) has the correct sample sequence but is compressed at the axis end. Reproduced with permission from Gauch (1982, Figure 3.15).

Numerous comparisons of popular algorithms have failed to produce a consensus regarding an optimal approach (e.g., Pielou 1984;
Minchin 1987; Wartenberg *et al.* 1987; Peet *et al.* 1988; Jackson and Somers
1991; van Groenewoud 1992; Palmer 1993; Økland 1996). In fact, PCA,
which was developed by Pearson in 1901, remains very popular among
ecologists. Widespread use of PCA is analogous to that of Simpson's index
and the Shannon–Weaver index, and may result from similar factors: (1)
numerous comparative studies illustrate that all algorithms have disadvantages; (2) ordination provides a quantitative, objective description of
community structure, but it does not provide information about the function or effective management of communities; and (3) PCA provides a suitable basis for the description and comparison of communities (Box 3.3).

PRINCIPAL COMPONENTS ANALYSIS

PCA was described by Pearson in 1901, but was largely ignored for three
decades (Hotelling 1933). In the 1930s, some psychologists were seeking
a single measure of intelligence, and reducing information to one axis,
or even a few axes, had considerable appeal. It was widely recognized
that some scores from standardized examinations were highly correlated (e.g., math and science), and PCA was proposed as a technique for
evaluating intelligence. Similarly, ecologists recognize that responses
of many species to environmental gradients are highly correlated.
Thus, the distribution or abundance of one species may explain the
distribution or abundance of many other species, and a few underlying
factors (e.g., environmental gradients) may be used to predict community structure.

Consider a simple, two-species community that is sampled with
three quadrats. Abundance of species X and Y is (3, 3), (4, 1), and (1, 5),
respectively, in quadrats 1, 2, and 3. Thus, summary statistics can be
calculated:

$$n_X = n_Y = 3 \qquad \sum X = 8 \qquad \text{mean of X} = 2.667 \qquad \sum X^2 = 26$$
$$\sum XY = 18 \qquad \sum Y = 9 \qquad \text{mean of Y} = 3 \qquad \sum Y^2 = 35.$$

Associated sums of squares corrected for the mean are:

$$\sum x^2 = 4.667 \qquad s^2_x = 2.333$$
$$\sum y^2 = 8 \qquad s^2_y = 4.$$

Values for species abundance can be plotted in quadrat-dimensional
space (Figure 3.9). Note that the length of the vector between the origin
and X is $|A| = \sum X^2 = 26$, and that the length of the vector between the ori-
gin and Y is $|B| = \sum Y^2 = 35$. Thus, the sums of squares have a direct geo-
metric interpretation. Furthermore, the angle between these two vectors

Box 3.3 Ordination as a management tool

By illustrating species composition in a simplified and straight-forward manner, ordination may facilitate management in a variety of ways. McPherson *et al.* (1991) attributed variation in herbaceous species composition, as reflected by reciprocal averaging ordination, to differences in site history and distance from trees in a juniper savanna. Specifically, long-term livestock grazing altered species composition, such that the first axis clearly separated herbaceous vegetation on a previously grazed site (Figure 3.8, bold type) from herbaceous vegetation on a relict site (plain type). The second ordination axis reflects differences in species composition attributable to distance from juniper plants: quadrats beneath woody plants (1) are differentiated from those at the canopy edge (2) or further from the juniper plant (3, 4, 5, and I are 1, 2, 3, and >5 m from the canopy edge, respectively). These short-statured (1–4 m tall) but dense-canopied woody plants offer physical protection from livestock and produce distinctive microenvironments, both of which may contribute to differences in herbaceous species composition. Further, there is no discernible pattern in herbaceous vegetation beyond the canopy edge, apparently because extensive juniper root systems extend throughout both sites.

Consistent with a large body of literature from these and other systems, long-term livestock grazing produced substantial and persistent effects on herbaceous species composition. These effects were evident at the scales of landscapes, communities, and individual plants. In addition, juniper plants protected some herbaceous species from livestock grazing, and favored shade-tolerant species capable of growing in a thick layer of litter. Finally, the influence of juniper plants on herbaceous vegetation extended at least 5 m beyond the canopy.

The importance of these findings depends on management goals. Livestock grazing and juniper plants affected the herbaceous plant community in a relatively complex manner, as evinced by a relatively simple ordination diagram. To the extent that these impacts on species composition influence management goals, managers may want to alter livestock grazing practices and manipulate the density of juniper plants.

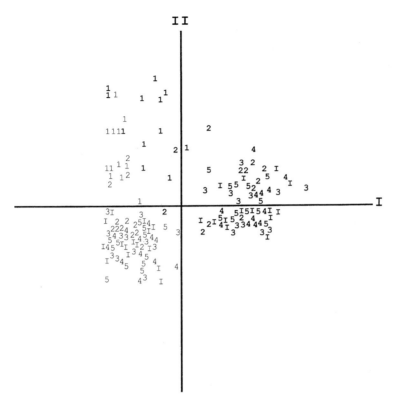

Figure 3.8 Reciprocal averaging quadrat ordination around *Juniperus pinchotii* trees on a semi-arid savanna grazed by livestock (bold type) and a nearby relict savanna site (normal type). Sampling location 1 is at the midpoint between tree bole and canopy edge, and is therefore beneath the canopy. Location 2 is at the canopy edge; and locations 3, 4, and 5 are 1, 2, and 3 m from the canopy edge, respectively. Location I is at least 5 m from the nearest tree. Reproduced with permission from McPherson *et al.* (1991).

is described by a simple relationship between the sums of squares and cross-products of the two species:

$|A||B| \cos\Theta = \Sigma XY.$
Solving for Θ,
$\cos\Theta = \Sigma XY/[(\Sigma X^2)(\Sigma Y^2)] = 18/(26)(35) = 0.60,$ so that
$\Theta = 53.4°.$

It is customary to work with corrected sums of squares (i.e., with data that are adjusted to a mean of 0). This is termed "centering" and produces the following adjusted values of abundance: $(X', Y') = (0.333, 0), (1.333, -2),$ and

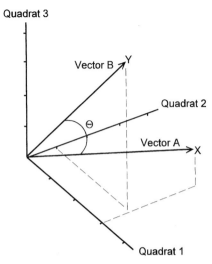

Figure 3.9 Two species (X and Y) plotted in quadrat-dimensional space. The abundance of species X is 3, 4, and 1 in quadrats 1, 2, and 3, respectively; the abundance of species Y is 3, 1, and 5 in quadrats 1, 2, and 3, respectively. The angle Θ between vectors A and B is 53.4°.

$(-1.667, 2)$, respectively, in quadrats 1, 2, and 3. Summary statistics are:

$$n_{X'} = n_{Y'} = 3 \qquad \sum X' = 0 \qquad \text{mean of } X' = 0 \qquad \sum X'^2 = 4.667$$
$$\sum X'Y' = -6 \qquad \sum Y' = 0 \qquad \text{mean of } Y' = 0 \qquad \sum Y'^2 = 8.$$

Associated sums of squares corrected for the mean are:

$$\sum x'^2 = 4.667 \qquad s^2_{x'} = 2.333$$
$$\sum y'^2 = 8 \qquad s^2_{y'} = 4.$$

Plotting these data is analogous to moving the coordinate system so that it is centered within the data (Figure 3.10). With the centered data, the length of the vector between the origin and X' is $|A'| = \sum X'^2 = 4.667$ and the length of the vector between the origin and Y' is $|B'| = \sum Y'^2 = 8$. The angle between these two vectors is:

$$|A'||B'|\cos\Theta' = \sum X'Y'.$$

Solving for Θ',

$$\cos\Theta' = -6/(4.667)(8) = -0.98, \quad \text{so that}$$
$$\Theta' = 169°.$$

Thus, with data that have been centered, the cosine of the angle between the two vectors is a correlation coefficient. For these data, species X and species Y are nearly perfectly negatively correlated. Uncorrelated species

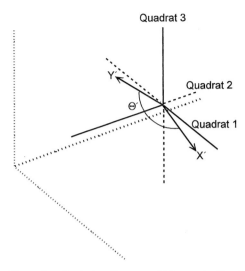

Figure 3.10 Centering the axes within species X and Y changes the coordinate system, but does not change the spatial relationships between species. The angle Θ' between the species is 169°, and the cosine of this angle is the correlation between the species ($r = -0.98$).

plot at right angles ($\cos 90° = 0$), and perfectly correlated species plot at an angle of 0° (perfectly positively correlated, $r = \cos 0° = 1$) or 180° (perfectly negatively correlated, $r = \cos 180° = -1$). PCA with centered data is conducted on a variance-covariance matrix, which assigns "weight" to species on the basis of abundance. It should be noted that centering does not change the relative positions of points: species X and Y are the same distance apart as "species" X' and Y'.

In addition to adjusting to a mean of 0 (i.e., centering the data), data may be standardized. This transformation merely calculates Z-scores by dividing centered data by standard deviations: $(X'', Y'') = (0.218, 0)$, $(0.873, -1)$, and $(-1.091, 1)$ respectively, in quadrats 1, 2, and 3. Summary statistics are:

$$n_{X''} = n_{Y''} = 3 \qquad \sum X'' = 0 \qquad \text{mean of } X'' = 0 \qquad \sum X''^2 = 2$$
$$\sum X''Y'' = -1.964 \qquad \sum Y'' = 0 \qquad \text{mean of } Y'' = 0 \qquad \sum Y''^2 = 2.$$

Associated sums of squares corrected for the mean are:

$$\sum x''^2 = 0 \qquad s^2_{x''} = 1$$
$$\sum y''^2 = 0 \qquad s^2_{y''} = 1.$$

With standardized data, species are equidistant from the origin. In this case, the $|A''| = |B''| = 2$. Therefore, standardization assigns all species equal "weight" in the ordination. PCA with centered and standardized data is conducted on a correlation matrix, which assigns equal "weight" to species, regardless of abundance. Unlike centering, standardization changes the relative positions of points in the data.

Because data are standardized to unit variance, PCA on a correlation matrix can accommodate data that are expressed in different units (i.e., are noncommensurate). Thus, environmental data can be included in the data set when conducting PCA on the correlation matrix, but not when conducting PCA on the variance–covariance matrix (i.e., on unstandardized data).

Deciding whether to conduct PCA on the variance–covariance matrix or on the correlation matrix has important consequences with respect to subsequent interpretability. These consequences are not always appreciated by ecologists or managers. For example, Rexstad *et al.* (1988) used a nonsensical data set comprised of noncommensurate data (e.g., meat prices, package weights of hamburger, book pages, random digits) as the basis for conducting PCA on the correlation matrix and on the variance–covariance matrix. PCA on the correlation matrix produced the expected uninterpretable result, with no variable accounting for more than 15% of the variability in the data. Two principal components explained over 99% of the total variance when the variance–covariance matrix was used; however, it would be inappropriate to consider these results meaningful because of the noncommensurate nature of the data (Taylor 1990). When units are commensurate and ordination is used as an exploratory tool (e.g., to generate hypotheses), it is appropriate to conduct PCA on both matrices and to interpret the results accordingly.

After data are centered and possibly standardized (depending on the nature of the data and preferences of the investigator), a "best-fit" line is drawn through the data. This line is the first principal component. The criterion for best fit is minimization of the sum of squares of perpendicular distances from the line to the points (i.e., all (X', Y') or (X'', Y'')). Axes are rotated so that the first axis (i.e., the first principal component) is horizontal on the page: this is termed "rigid rotation." Species are then plotted in the new coordinate system (Figure 3.11). Subsequent lines of best fit are projected through the data, subject to the constraint that all such lines are uncorrelated with previous lines. These lines represent principal components 2, 3, and so forth, up to a maximum of $n-1$ axes, where $n =$ the number of quadrats.

Rigid rotation does not change the relative positions of points, nor does it alter the total variability (which is sometimes termed "dispersion") in the data. However, rigid rotation increases the proportion

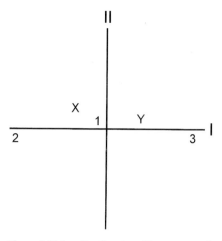

Figure 3.11 Results of centered but not standardized principal components analysis for species X and Y with abundances of (3, 3), (4, 1), and (1, 5) in quadrats 1, 2, and 3, respectively.

of variability associated with the first axis. This is an important characteristic when analysis involves many species or quadrats, which often occurs in natural communities.

The geometric interpretation of the PCA algorithm can be summarized as follows. Species are plotted in quadrat-dimensional space. Data are centered, and possibly standardized (depending on the nature of the data and preferences of the investigator). A "best-fit" line is projected through the data. Subsequent "best-fit" lines are drawn through the data, subject to the constraint of orthogonality between axes.

PCA axes concentrate variance of the point configuration into relatively few axes, in contrast to the high dimensionality of the original data. In effect, PCA summarizes complex data into a form that may be comprehended, and therefore interpreted. However, the ecological interpretation of PCA axes is necessarily a subjective exercise. Ideally, few axes (components) can be interpreted in light of species' natural histories and also explain considerable variability in the data. The amount of variability explained by the first few axes is less important than the ability to interpret the axes with respect to the natural history of the species. For example, the ability of the first few axes to explain variability may be associated with results that are ecologically meaningless or severely distorted (as was the case with PCA on the variance–covariance matrix with noncommensurate data: Rexstad *et al.* 1988; Taylor 1990). Conversely, the first few axes may explain little of the total variance and yet be ecologically informative.

BACK TO THE FUTURE: THE REBIRTH OF DIRECT GRADIENT ANALYSIS

Canonical correspondence analysis (CCA) represents an extension of RA (Jongman *et al.* 1987) which was "developed to relate community composition to known variation in the environment" (ter Braak 1986). CCA is available in the CANOCO computer program (ter Braak 1988), and is "becoming the most widely used gradient analysis technique in ecology" (Palmer 1993:2215). CCA explicitly incorporates environmental data into the ordination algorithm, and is therefore a form of DGA.

CCA and ordination are conceptually different, yet they appear to be similar (Økland 1996). Both give rise to plots of species and/or samples with respect to axes that are routinely called "ordination diagrams" or simply "ordinations." Furthermore, the two approaches are apparently used interchangeably by many ecologists. Finally, most authors do not justify their choice of algorithms, which further contributes to the idea that CCA is equivalent to ordination. The recent development and widespread use of CCA, coupled with accolades from mathematical ecologists (e.g., ter Braak and Prentice 1988; Palmer 1993), give the impression that CCA is superior to ordination.

As with most other activities, goals and objectives should motivate the selection of a multivariate algorithm; the advantages and disadvantages of various techniques should be appropriately appreciated and considered. Økland (1996) demonstrated that CCA (which he included in the broad category of "constrained ordination") and ordination serve different purposes in ecological analyses. Specifically, ordination is particularly useful for approximating community structure and generating hypotheses about underlying environmental gradients. Explicit incorporation of environmental variables in CCA may mask important gradients in species composition, thereby inhibiting the generation of hypotheses (Økland 1996). However, use of CCA is appropriate to describe how species respond to specific observed environmental variables (McCune 1997), and CCA provides an estimate of the statistical significance of ordination axes or environmental variables, if data meet the assumptions of statistical tests (Økland 1996).

Classification

Classification is used to identify members of discrete classes. As such, classification assumes that groups are present in the data, but that relationships between quadrats have not been elucidated. Several general approaches can be used to classify vegetation. Cluster analysis represents an objective and repeatable numerical approach that is employed frequently enough to merit discussion in this chapter.

Similar to ordination, cluster analysis uses species-abundance (or species-occurrence) data to determine relationships between samples (i.e., quadrats). It is typically used to display patterns of community structure and it, therefore, serves as a basis for the generation of hypotheses about factors underlying community structure. In addition, cluster analysis may be used to develop vegetation maps.

The specific objective of cluster analysis is to place similar samples into groups. Criteria for similarity among samples and size of groups have been proposed, debated, and reviewed (e.g., Milligan and Cooper 1985; Casado *et al.* 1997). The relative utility of various "stopping rules" depends on the properties of the underlying data, and no rule has emerged as superior in most cases. Therefore, the most widely used method involves subjective selection of group size and number (the "phenon line" of Sneath and Sokal 1973).

Several specific algorithms have been developed to conduct cluster analysis. Interpretation of results is broadly similar among these techniques. Therefore, this section will focus on the interpretation of results derived from cluster analysis; the following section then describes several of the common clustering algorithms. A geometric approach is used throughout, analogous to the description of PCA in the previous section. More formal mathematical treatments of cluster analysis and other classification approaches are provided by several authors (e.g., Pielou 1977; Gauch 1982; Romesburg 1984; Manly 1986; Jongman *et al.* 1987).

Results of cluster analysis are usually displayed with dendrograms, which provide a simple visual summary of clustering algorithms. Individual samples are displayed along the horizontal axis, and distance between samples is displayed on the vertical axis. Thus, similar samples are joined by short vertical lines and they usually occur near each other. Dissimilar samples are joined by long vertical lines, and are usually far apart and joined to other samples or groups first. Clustering algorithms typically display relationships among samples, but not among species.

The distance between samples and the order in which they are joined together provide the basis for classification. The distance between groups or the number of groups is specified; this step is ultimately arbitrary and subjective. If distance is used as the criterion for classification, then samples which join at a distance shorter than the one specified are said to belong to the same group. If number of groups is used as the criterion for classification, then a horizontal line is drawn directly above the specified number of groups: samples that are joined together below this line are said to belong to the same group. As with ordination, there is no guarantee that these criteria will shed appreciable light on community structure or produce meaningful categories for classification, because cluster analysis is a mathematical exercise: it is not designed to recognize and integrate ecological information. Obviously, results of cluster analysis do reflect relationships between samples most of the time, as evidenced by the widespread use of clustering algorithms by ecologists (Box 3.4).

In addition to the use of cluster analysis as a basis for vegetation classification, dendrograms can be used to generate hypotheses about community structure. Several quantitative techniques may facilitate the generation of these hypotheses. For example, descriptive statistics can be calculated at each dichotomy of interest. Means and standard deviations of the environmental variables and the frequency or abundance of various species may suggest a link between a specific environmental factor and the distribution or abundance of a species. In addition, discriminant analysis (Lachenbruch 1975) may be conducted at each dichotomy to identify quickly environmental variables that are particularly disparate on either side of the dichotomy (discriminant analysis is a regression-like

---

Box 3.4 Cluster analysis as a management tool

Cluster analyses indicate the community-level similarity of communities with each other, and therefore facilitate management by furthering communication and by assessing the uniqueness of specific communities. Mehlert and McPherson (1996) used minimum-variance clustering (Ward 1963) to describe the number and character of communities within the broadly defined oak woodland vegetation type of the southwestern United States. Oak woodlands had not been studied in sufficient detail to allow the identification of communities at a level of resolution suitable for management.

The southwestern oak woodland was classified into 15 distinct communities (Figure 3.12). The dominant two communities comprised about 30% of all oak woodlands in the region. In contrast, two communities were represented by only five of 374 quadrats (1.3%). If management goals include conservation of communities, efforts should focus on these relatively rare communities. However, research directed at the response of these communities to ongoing resource extraction, such as fuelwood harvest and livestock grazing, should focus on the widespread communities first. Given that the two dominant communities are among the most dissimilar communities, research results from these communities may be particularly revealing and may be generally applicable to the remaining communities.

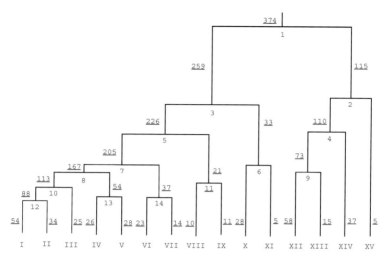

Figure 3.12 Dendrogram of oak woodland communities in the southwestern United States. Communities were defined with minimum-variance clustering (Ward 1963) of 374 plots, using basal area of woody plant taxa as a measure of abundance. Arabic numerals label dichotomies, underlined arabic numerals are numbers of plots, and roman numerals indicate communities. Groups represented by only one plot were regarded as outliers and are therefore not shown. Communities are described in Table 3.2. Reproduced with permission from Mehlert and McPherson (1996).

technique used to identify variables that maximally discriminate between two or more sets of data; Williams (1983) provides a review and critique). Finally, significance tests (e.g., *t*-tests) can be used to determine whether environmental variables are statistically different on each side of a dichotomy. Results of these tests should be interpreted with considerable caution, and they are appropriate only when the data meet specific assumptions. In fact, when cluster analysis is appropriately used as an exploratory tool for the generation of hypotheses, there is no need to conduct significance tests. After hypotheses are generated, experiments may be used to test for causal relationships between environmental variables and community structure.

Clustering algorithms

This section will consider a simple two-species community that has been sampled with six quadrats. Abundance of species A and B is (15, 9), (12, 8), (17, 13), (0, 7), (8, 0), and (3, 12), respectively, in quadrats 1 through to 6. These data can be plotted in species-dimensional space (Figure 3.13).

Table 3.2 *Communities within oak woodlands of the southwestern United States, defined with minimum-variance clustering (Ward 1963)*

Number	Community name	Elevation (m) Mean	Range	Basal area (m²/ha) Mean	Range	Quercus basal area (m²/ha) Mean	Range
I	Low elevation, low basal area mixed oak	1670	1160–2100	6.5	0.5–19.3	3.6	0.0–18.9
II	Low basal area *Quercus arizonica*	1720	1460–1980	12.9	4.7–26.9	7.2	1.9–12.6
III	*Q. emoryi*	1620	1340–1890	13.5	5.8–27.9	11.6	2.1–24.1
IV	*Quercus* spp. *Juniperus deppeana*	1800	1400–2440	20.1	7.9–35.4	9.1	4.8–17.1
V	Low basal area *J. deppeana*	1840	1710–2380	16.7	6.8–28.7	3.9	0.5–9.6
VI	*Pinus ponderosa Q. gambelii*	1940	1770–2160	23.2	15.6–43.7	7.9	1.8–21.9
VII	*P. ponderosa Q. arizonica*	1880	1800–1980	25.6	18.8–39.2	9.1	0.1–18.4
VIII	Very high basal area *J. deppeana*	1900	1710–2130	41.4	30.8–49.8	13.5	5.3–20.5
IX	*J. deppeana Q. arizonica*	1870	1770–1980	37.6	30.1–51.7	15.8	9.8–23.0
X	Medium basal area *Q. arizonica*	1810	1430–2260	29.2	12.8–53.8	24.5	12.8–42.2
XI	*Q. hypoleucoides*	1870	1620–2010	43.7	25.8–80.4	40.2	22.0–80.4
XII	Low basal area *Quercus* spp.	1740	1280–2380	18.0	6.2–51.6	11.7	0.0–19.1
XIII	High basal area *Quercus* spp.	1740	1370–2100	35.5	24.9–47.0	26.8	20.0–36.6
XIV	High elevation *P. ponderosa Quercus* spp.	2020	1740–2320	26.4	7.9–50.0	8.9	0.0–26.2
XV	Very high basal area *Quercus* spp.	1790	1680–2010	65.2	47.2–80.8	53.7	44.2–66.5

Roman numerals refer to communities in Figure 3.12.

Source: Modified with permission from Mehlert and McPherson (1996).

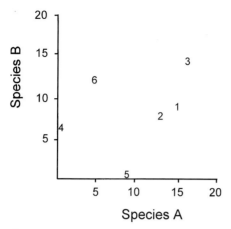

Figure 3.13 Six quadrats plotted in species-dimensional space. Abundance of species A and B is (15, 9), (12, 8), (17, 13), (0, 7), (8, 0), and (3, 12), respectively, in quadrats 1 through to 6.

At this point, two distinct approaches can be used to form groups from these data: agglomerative and divisive. Agglomerative algorithms begin by treating each sample as a group, then forming a group from the most similar pair of samples, and so on until all samples are contained in one group. Divisive algorithms begin by treating the entire set of samples as a group, then dividing this group into two groups, and so on until each sample is its own group. Most clustering algorithms that are widely used by ecologists are agglomerative. Exceptions include two-way indicator species analysis (TWINSPAN) (Hill *et al.* 1975; Hill 1979) and a closely related algorithm, constrained indicator species analysis (COINSPAN) (Carleton *et al.* 1996).

Single-linkage clustering (Sneath and Sokal 1973), which is also termed nearest-neighbor clustering, is the oldest and simplest clustering algorithm. Although single-linkage clustering is rarely used by contemporary ecologists, its simplicity serves as a convenient starting point for a discussion of clustering algorithms. Single-linkage clustering employs Euclidean distance as the measure of similarity between groups, and the distance is defined as the shortest distance involved in a comparison of two groups. Euclidean distance is defined as:

$d\,(j, k) = \sum\sum (N_{ij} - N_{ik})^2$,   where

$d\,(j, k)$ = distance between samples j and k,

$\quad N$ = abundance of a species in a specified sample,

$\quad i$ = specified species,   and

$\quad s$ = total number of species.

From our example with six samples and two species, the Euclidean distance between samples 1 and 2 is:

$$d(1,2) = \sqrt{(15-12)^2 + (9-8)^2} = \sqrt{10} = 3.16.$$

Furthermore, all possible distances between points (i.e., samples) can be calculated:

	1	2	3	4	5	6
1	—	3.16	4.47	15.16	11.40	12.32
2		—	7.07	12.04	8.94	9.85
3			—	18.03	15.81	14.04
4				—	10.63	5.83
5					—	13.00

The smallest number in this matrix is 3.16, which indicates that samples 1 and 2 are nearer than any other pair of samples. Single-linkage clustering (and most other agglomerative algorithms) therefore joins samples 1 and 2 into a group. Note that single-linkage clustering ignores the properties of the groups: only individual samples are compared. All possible distances between entities are calculated, with distance defined as the shortest distance involved in a comparison of two entities (e.g., the distance between samples 1 and 3 is 4.47 and the distance between samples 2 and 3 is 7.07, so the distance is defined as the shorter of the distances, 4.47):

	1, 2	3	4	5	6
1,2	—	4.47	12.04	8.94	9.85
3		—	18.03	15.81	14.04
4			—	10.63	5.83
5				—	13.00

The smallest number in this matrix is 4.47, which indicates that sample 3 joins the previously formed group. The subsequent matrix is:

	1,2,3	4	5	6
1,2,3	—	12.04	8.94	9.85
4		—	10.63	5.83
5			—	13.00

The smallest number in this matrix is 5.83, which indicates that samples 4 and 6 join to form a group. This leaves the following distances between entities:

	1,2,3	4,6	5
1,2,3	—	9.85	8.94
4,6		—	10.63

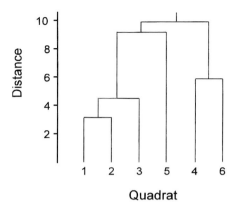

Figure 3.14 Dendrogram for the data shown in Figure 3.13, based on single-linkage clustering.

The smallest number in this matrix is 8.94, which indicates that sample 5 joins the group comprised of quadrats 1, 2, and 3. Finally, the two remaining groups are joined at a distance of 9.85. The resulting dendrogram summarizes the algorithm (Figure 3.14).

Single-linkage clustering is rarely used to analyze ecological data. This algorithm ignores group properties by calculating distances based only on the distances between individual samples. In addition, single-linkage clustering is said to be "space contracting:" as a group grows, it becomes more similar to other groups. The space-contracting nature of single-linkage clustering tends to cause samples to be added to preceding groups one at a time (this is sometimes termed "chaining"), which hampers interpretability of the resulting dendrogram.

Complete-linkage (farthest-neighbor) clustering (Sneath and Sokal 1973) is identical to single-linkage clustering except that the distance between entities is defined as the point of maximum distance between samples in the groups being compared. For example, the distance between group (1,2) and sample 3 is the maximum distance involved in the comparison (7.07) rather than the minimum distance used in single-linkage clustering (4.47).

Complete-linkage clustering overcomes the "chaining" produced by single-linking clustering. In fact, complete-linkage clustering is "space dilating:" as a group grows, it becomes less similar to other groups. As with single-linkage clustering, group properties are ignored because distances are calculated on the basis of distance between individual samples.

Average-linkage clustering and centroid clustering (Sokal and Michener 1958) are similar to single-linkage and complete-linkage

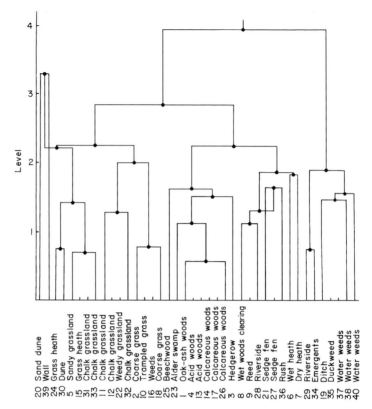

Figure 3.15 Dendrogram with reversals. Reproduced with permission from Gauch and Whittaker (1981).

clustering, except that average distances form the basis for joining entities. Therefore, the properties of groups are used to assess similarity, an approach which is viewed as advantageous relative to single-linkage and complete-linkage clustering. However, these techniques are characterized by peculiar behavior that creates difficulties with interpretation and which, subsequently, has limited their widespread adoption. Specifically, these algorithms have the potential to exhibit "reversals," a situation in which entities are joined at a shorter distance than a previous fusion (Figure 3.15). This implies that entities joined in the second fusion are more similar than those joined in the first.

Single-linkage, complete-linkage, average-linkage, and centroid clustering optimize the route along which structure is sought in the data. These algorithms focus on the distance between entities, not on the properties of the entities themselves. These algorithms are within the general class of techniques called hierarchical methods. In contrast, nonhierarchical methods optimize some property of the group being formed (e.g., increase in sums of squares). Thus, group properties are the explicit focus of nonhierarchical methods.

Minimum-variance clustering (syn. Ward's method, Orlóci's method) (Ward 1963) is a nonhierarchical clustering algorithm that uses the properties of groups to assess similarity. Thus, it incorporates information about groups, not merely about individual samples. Like the clustering algorithms previously described, minimum-variance clustering employs Euclidean distance as a distance measure. The concept underlying minimum-variance clustering is that the distance between members of a group and the group's centroid (i.e., center) can be used as an indicator of group heterogeneity. Specifically, the fusion rule used by minimum-variance clustering is: join groups only if the increase in the squared distances is less for that pair of groups than for any other pair. Thus, minimum-variance clustering minimizes heterogeneity within groups and therefore favors the formation of small clusters of approximately equal size.

This algorithm lends itself to a measure of classification efficiency; the total sums of squares is expressed as the squared distance between all quadrats and the centroid – in our example, the group centroid is given by $((15+12+17+0+8+3)/6, (9+8+13+7+0+12)/6)$, or (9.167, 8.167). At any point in the analysis, sums of squares can be calculated for each group, and this measure represents within-group heterogeneity. The proportion of total variability explained by a particular group is an indicator of that cluster's importance in the data set (i.e., SS_{group}/SS_{total}).

TWINSPAN is a divisive clustering method that relies on an ordination algorithm, RA. A crude dichotomy is formed in the data, with the RA centroid serving as the dividing line between two groups. This dichotomy is refined by a process comparable to iterative character weighting (Hogeweg 1976), a summary of which is provided by Jongman *et al.* (1987:194–5). Dichotomies are then "ordered" so that similar clusters are near each other. The TWINSPAN algorithm ensures that dichotomies are determined by relatively large groups, so these dichotomies depend on general relations rather than on single observations, which may be atypical.

Unlike prior clustering algorithms, TWINSPAN also produces a classification of species. This classification is based on the fidelity of species to specific samples or clusters of samples. Thus, in addition to a dendrogram, a structured table is produced (Table 3.3). However, a structured table produced from large data sets is rarely presented because patterns in a large table are not readily discernible.

Table 3.3 *Structured table from TWINSPAN*

No.	Species	11	17	18	19	5	6	7	10	1	2	3	4	8	9	12	13	14	15	16	20	Dichotomy
3	Air pra	.	2	.	3	.	.	.	.	.	.	.	.	.	.	.	.	.	.	.	.	00000
12	Emp nig	.	.	.	2	.	.	.	.	.	.	.	.	.	.	.	.	.	.	.	.	00000
13	Hyp rad	2	2	.	5	.	.	.	.	.	.	.	.	.	.	.	.	.	.	.	.	00000
28	Vic lat	2	.	1	.	.	.	.	1	.	.	.	.	.	.	.	.	.	.	.	.	00000
5	Ant odo	.	4	.	4	4	3	2	4	.	.	.	.	.	.	.	.	.	.	.	.	00001
18	Pla lan	3	2	3	.	5	5	5	3	.	.	.	.	.	.	.	.	.	.	.	.	00010
1	Ach mil	.	2	.	.	2	2	2	4	1	3	.	.	.	.	.	.	.	.	.	.	000110
26	Tri pra	.	.	.	.	2	5	2	.	.	.	.	.	.	.	.	.	.	.	.	.	000110
6	Bel per	.	.	2	.	2	.	.	2	.	3	2	2	.	.	.	.	.	.	.	.	000111
7	Bro hor	.	.	.	.	2	.	2	4	.	4	.	3	.	.	.	.	.	.	.	.	000111
9	Cir arv	.	.	.	.	.	.	.	.	.	.	.	2	.	.	.	.	.	.	.	.	000111
11	Ely rep	.	.	.	.	4	.	.	.	4	4	4	4	.	6	.	.	.	.	.	.	001
17	Lol per	7	.	2	.	2	6	6	6	7	5	6	5	4	2	.	.	.	.	.	.	001
19	Poa pra	4	1	3	.	2	3	4	4	4	4	5	4	4	4	.	2	.	.	.	.	001
23	Rum ace	.	.	.	.	5	6	3	.	.	.	.	.	.	2	2	.	.	.	.	.	001
16	Leo aut	5	2	5	6	3	3	3	3	.	5	2	2	3	2	2	2	2	2	.	2	01
20	Poa tri	.	.	.	.	6	4	5	4	2	7	6	5	4	5	4	9	.	.	2	.	01
27	Tri rep	3	.	2	2	2	5	2	6	.	5	2	1	2	3	3	2	6	1	.	.	01
29	Bra rut	4	.	6	3	2	6	2	2	.	.	2	2	2	2	4	.	.	4	4	4	01
4	Alo gen	.	.	.	.	.	.	.	.	.	2	7	2	5	3	8	5	.	.	4	.	10
24	Sag pro	2	.	.	3	.	.	.	.	.	.	.	5	2	2	4	2	.	.	.	.	10
25	Sal rep	.	.	3	3	.	.	.	.	.	.	.	.	.	.	.	.	.	.	.	5	10
2	Agr sto	.	.	.	.	.	.	.	.	.	.	4	8	4	3	4	5	4	4	7	5	110
10	Ele pal	.	.	.	.	.	.	.	.	.	.	.	.	4	.	.	.	4	5	8	4	11100
21	Pot pal	.	.	.	.	.	.	.	.	.	.	.	.	.	.	.	.	2	2	.	.	11100
22	Ran faa	.	.	.	.	.	.	.	.	.	.	.	2	.	.	2	2	2	2	.	4	11100
30	Cal cus	.	.	.	.	.	.	.	.	.	.	.	.	.	.	.	.	4	.	3	3	11100
14	Jun art	.	.	.	.	.	.	.	.	.	.	.	4	4	.	.	.	.	3	3	4	11101
8	Che alb	.	.	.	.	.	.	.	.	.	.	.	.	.	.	.	.	1	.	.	.	1111
15	Jun buf	.	.	.	.	.	.	2	.	.	.	.	.	.	.	.	.	4	4	3	.	1111
		0	0	0	0	0	0	0	0	0	0	0	0	1	1	1	1	1	1	1	1	
		0	0	0	0	1	1	1	1	1	1	1	1	0	0	0	0	1	1	1	1	
						0	0	0	0	1	1	1	1									

Numbers for species (rows) represent abundance in quadrats (columns); a dot indicates that the species was absent from the quadrat. Zeros and ones on the right-hand side and the bottom of the table indicate the dichotomies.

Source: Modified with permission from Jongman *et al.* (1987)

Minimum-variance clustering and TWINSPAN are computationally complex and time consuming compared with clustering algorithms developed earlier. However, the widespread availability of powerful and inexpensive computers during the last 15 years has allowed them to become the most widely used clustering algorithms in ecological studies.

Process models

There are three primary process-oriented models of plant community structure (Austin 1986; Keddy and MacLellan 1990): the plant strategy model (Grime 1979), the gap dynamics/regeneration model (Grubb 1977; Pickett and White 1985), and the resource-ratio hypothesis (Tilman 1985, 1988). These models have not been applied to the management of plant communities, at least partially because they are not yet adequately integrated with descriptive approaches (Keddy and MacLellan 1990). Thus, they consider the effects of resource availability and disturbance on the interactions between plants, but they do not predict species composition in complex communities.

An additional model of community structure, the centrifugal organization model (Rosenzweig and Abramsky 1986; Keddy 1989, 1990), incorporates underlying mechanisms into descriptions of community structure along environmental gradients. Thus, this model integrates pattern with process. The centrifugal organization model represents an extension of the competitive hierarchy model described in the previous chapter. The model assumes that all species share a central habitat in which they exhibit maximum performance (e.g., growth, survival, reproductive output), but that each species has another (peripheral) habitat in which it is the best competitor. This model represents a variant of inclusive niche structure, in which species have overlapping fundamental niches along only one axis. Entire environmental gradients ("niche axes") may radiate outward from the central preferred habitat in the centrifugal organization model. Near the center (i.e., optimal habitat), species may have entirely inclusive fundamental niches; at or near the periphery, species' fundamental niches may include only a few adjacent neighboring species in the direction of the central habitat. In the latter case, negative interactions would be completely asymmetrical (i.e., interference), and removal experiments should show a species to increase nearer the central habitat, but not toward the periphery. Wetlands appear to be organized in this manner: a central habitat characterized by low disturbance and high fertility is dominated by large leafy species capable of forming dense canopies (e.g., *Typha* spp.), constraints such as

Figure 3.16 Densely growing cattail dominates the wetland community of the Ciénega de Santa Clara near the mouth of the Colorado River, Sonora, Mexico. Photo by Stephen DeStefano.

disturbance and fertility create radiating axes along which different groups of species and vegetation types are arrayed, and rare species occur only in peripheral habitats with low biomass (Moore and Keddy 1989; Moore *et al.* 1989; Keddy 1990) (Figure 3.16). Communities of forest trees (Keddy and MacLellan 1990) and desert rodents (Rosenzweig and Abramsky 1986) also exhibit distribution patterns consistent with the centrifugal organization model.

SUMMARY

The techniques outlined in this chapter can be applied to assemblages of organisms, and have been effectively applied to animal and plant communities. These techniques enable the objective description of communities, and thus serve as an important link between ecological science and natural resource management. In addition, effective description of communities facilitates the comparison of vegetation or animal communities on different sites or at different times. As such, the tools presented in this chapter allow the objective assessment of management strategies or environmental change. These approaches form a foundation for evaluating and interpreting community change over time.

4

Succession

Although ecosystems may be sufficiently stable to allow objective characterization, as described in the previous chapter, they are not temporally static entities. Rather, they are characterized by changes in species composition at various temporal scales. This chapter will focus on the factors that underlie transitions from one state to another at the temporal scales of years to decades. Changes in ecosystems at these temporal scales are termed succession (Figure 4.1).

Identifying states of ecosystems and determining causes of transitions from one state to another are fundamental to effective management. Specifically, management efforts should aim to achieve states that meet clearly specified objectives. In addition, managers must be cognizant of states that are not feasibly attainable and set management expectations accordingly.

It is also important to remember that wildlife communities are greatly influenced by vegetation succession. This relationship has been well documented in forests, but it also applies to other ecosystems, such as wetlands and grasslands. For example, Lloyd *et al.* (1998) found that changes in woody plant structure, primarily invasion of mesquite attributed to fire suppression, contributed to changes in the composition of grassland bird communities in former grasslands of southern Arizona. Managers should also recognize that habitat "quality" for many wildlife species can change with vegetation succession; good (or poor) quality habitat at one point in time often changes within a few years or decades (see below for a discussion of "habitat quality" and "habitat fitness"). Thus, it is critical that wildlife managers understand the major models of vegetation succession developed and debated by plant ecologists, and that they appreciate the implications of differing viewpoints to the understanding and management of plant, and thus animal, communities.

Figure 4.1 In New England, old-field succession is the often used example to represent classic succession: abandoned farm fields revert to herbaceous plants, which are replaced by shade-intolerant woody growth and then by shade-tolerant trees. In southern New England, the so-called "climax" forest is referred to as birch–beech–maple (yellow birch, American beech, and sugar maple). Photo by Stephen DeStefano.

After describing various models of succession, latter sections of this chapter will discuss tools to study succession. Once again, the focus is on plant ecology and how plant ecologists might approach research into the processes and mechanisms of vegetation succession. Wildlife biologists can apply these same approaches – i.e., retrospective studies, monitoring, comparisons, experiments – to animal populations and the influence of vegetation succession on the distribution and structure of animal communities.

TRADITIONAL VIEW

General patterns of ecosystem change were acknowledged by early Roman writers and were described by many naturalists in the 18th and 19th centuries (Spurr 1952). The earliest scientific work was conducted in sand dunes by Cowles (1899), and patterns described by naturalists and Cowles were formalized by Clements in a series of papers during the early part of the 20th century. Clements' 1916 book on succession was particularly influential in establishing a paradigm for ecosystem dynamics. Clements promoted the view that ecosystems are organic

Figure 4.2 Large-scale disturbances such as clearcutting not only alter vegetation structure and plant succession, but can dramatically affect hydrology, sediment transport, soil microfauna, microclimate, and other ecological processes. Photo by Stephen DeStefano.

units; in fact, the Clementsian view of ecosystem dynamics is often termed "organismic" and ecosystems or plant communities are called "superorganisms." The organismic concept is regarded by most contemporary ecologists as an inappropriate view (but see Wilson and Sober 1989; Wilson 1997). Nonetheless, this view remains the dominant paradigm promoted in ecology texts and it forms the basis for site classification in land management agencies (i.e., delineation of range sites and habitat types). The Clementsian view will, therefore, be briefly described (Figure 4.2).

Clements (1916) described succession as a sequence of identifiable stages, which he termed nudation, ecesis, competition, reaction, and stabilization. Nudation is the process which creates a patch of bare soil, and it is said to initiate succession. Ecesis is the successful establishment of plants, coming either from propagules remaining in the soil (i.e., seeds, root fragments, or whole plants) or migrating from elsewhere. Ecesis is said to be controlled by environmental conditions and the characteristics of plant species available at the site. Competition among established plants then leads to the elimination of some species. Reaction is the change in the physical environment that results from the growth and death of plants, and it contributes to continual change of resource availability. Finally, stabilization occurs as long-lived species dominate a site.

Clements indicated that this phase rarely, if ever, occurs. It is synonymous with "climax."

Clements' conceptual model of succession is descriptive, but it is not explanatory or mechanistic. Thus, Clements simply formalized the process of species' replacement described by naturalists during the preceding two centuries.

The traditional view that developed as a result of Clements' work was that of "relay floristics." According to this view, species prepare an area to make it more suitable for other species. Because ecosystems and communities are superorganisms, they are capable of employing a strategy, including the strategy of site preparation. This idea has been repeated in dozens of papers and books, including relatively recent textbooks. For example, Odum (1983) discussed succession within a section titled "The Strategy of Ecosystem Development:"

> [S]uccession is an orderly process of community development; it is reasonably directional and, therefore, predictable . . . it results from modification of the physical environment by the community; that is, succession is community-controlled . . . Species replacement in the sere occurs because populations tend to modify the physical environment, making conditions favorable for other populations until an equilibrium between biotic and abiotic is achieved.

Thus, the prevailing paradigm in ecology from the early 1900s until relatively recently was that ecosystems facilitate the development of other ecosystems by altering site conditions. This implies that late-successional species could not occupy the site without earlier occupation by earlier-successional species.

This traditional Clementsian explanation of vegetation succession is the most familiar model to wildlife managers, and represents the only model with which many managers are familiar. Undergraduate wildlife biology curricula usually include only two to three botany courses, one to two of which are often plant taxonomy, with only one course in plant ecology. Thus, wildlife biology majors are not exposed to a deep understanding of the modern views of vegetation succession. Yet, vegetation manipulation via alteration of vegetation succession – using techniques such as timber harvesting, prescribed burning, plowing, or disking – to manipulate animal populations is a frequent approach used by wildlife managers. The following sections offer alternative views of succession; the underlying processes and mechanisms have numerous implications for the research, management, and conservation of plant and animal communities.

CHANGING VIEWS

Egler (1954) studied secondary succession in abandoned agricultural fields. He concluded that, when disturbed, the soil retains a large and diverse pool of propagules. These propagules represent various successional stages, so that virtually all species are present from the time of disturbance and different species assume dominance over time. Egler termed this conceptual model "initial floristics," indicating that all or nearly all species are present initially (as seeds or seedlings). If a species is absent from the propagule pool, it will not participate in succession or will do so only very slowly. According to this view, succession represents changes in the dominance of species over time.

Drury and Nisbet (1973) reviewed the field evidence for relay floristics and initial floristics, and reached several important conclusions. First, many species that characterize late-successional stages are present but inconspicuous at earlier stages. Second, removal of annual plants during the first few years of secondary succession often enhances the performance of perennials. Similarly, removal of early-successional pines in forests accelerates dominance by late-successional hardwoods. Third, most studies suggest that early-successional stages can be explained in terms of differential growth in response to changing resource availabilities. These changes in resource availabilities are a natural consequence of changes in species composition on a site. Finally, late-successional plants are usually present throughout the history of vegetation change, and early-successional plants often delay the rate of succession.

Drury and Nisbet concluded that there was considerably more support for the initial floristics model than for the relay floristics model. They viewed succession as a process in which plant species are sorted along a gradient of resources. Species replacement occurs because each individual species has a unique optimum for growth and reproduction and because resource availabilities change through time. Drury and Nisbet's ideas represented a fundamental shift from succession as a community-controlled phenomenon to a process based on the properties of individual species. Pickett (1976) expanded Drury and Nisbet's resource-gradient concept to include interference: species replacements occur during succession as a result of changes in competitive "winners" in a changing environment. The centrifugal organization model (Rosenzweig and Abramsky 1986; Keddy 1990; Wisheu and Keddy 1992) discussed in Chapter 3 formalizes the relationship between interactions, environment, and ecosystem structure.

Connell and Slatyer (1977) developed three conceptual models for succession, then reviewed the literature to determine the amount of evidence to support each model. According to their facilitation model (Model 1), colonists alter the environment and thereby make the site amenable to occupation by other species. Although Connell and Slatyer did not imply the existence of a "strategy" *vis-à-vis* Odum, this model is based on a Clementsian interpretation of succession. Connell and Slatyer concluded that Model 1 was associated with primary succession (e.g., when nitrogen-fixing species colonize glacial till or river sand and therefore contribute to soil enrichment). They attributed this model to relay floristics. According to the tolerance model (Model 2), environmental modifications imposed by early-successional species neither increase nor reduce the rates of recruitment and growth of later-successional species. Thus, species replacement patterns are solely dependent on life history: late-successional species occupy a site either early or late in the course of succession, but are able to grow slowly and reproduce despite the presence of early-successional species. Connell and Slatyer concluded that few situations in the literature fit this model, which they attributed to initial floristics. According to the inhibition model (Model 3), early colonists secure space and/or resources and then inhibit subsequent invasion by other species or suppress the growth of species that invade at the same time. On the death of an early colonist, space and/or resources are released for another individual (potentially of the same species) and this leads to succession. This model is driven by negative interactions between species, and was attributed by Connell and Slatyer to initial floristics.

By this point in time, ecologists were making a clear shift away from Clementsian-based explanations driven by vegetation *per se*. In addition, they were beginning to abandon models with universal utility, and were instead focusing on individual species. Models developed later were consistent with these trends.

Noble and Slatyer (1980) attempted to define the vital attributes of species that would allow prediction of their performance during succession. These vital attributes were based on three characteristics of plants: (1) method of arrival or persistence after disturbance; (2) ability to enter an existing ecosystem and then grow to maturity; and (3) time required to reach critical stages in the species' life cycle (e.g., reproductive maturity). Pickett *et al.* (1987) expanded the concept of vital attributes and developed a hierarchy of succession including causes of succession, contributing processes, and defining factors.

Vital attributes were precursors to assembly rules and response rules. Assembly rules (Diamond 1975) are used to predict species composition for a specified habitat from the total species pool in a region. They are based on the traits (i.e., attributes) of species. Response rules are derived from assembly rules, and are used to predict the changes in species composition that result from changes in environment or land use (Keddy 1989, 1992). Development of assembly rules and response rules grew out of the trait-based vital attributes approach of Noble and Slatyer, and these rules represent primary goals of community ecology (Keddy 1992); they are consistent with the state-and-transition model of succession (Westoby *et al.* 1989).

The state-and-transition model acknowledges that successional pathways can be complex and that these pathways do not necessarily converge on a single endpoint ("climax"). In particular, Westoby *et al.* recognized that the concept of single-equilibrium systems which progress steadily toward "climax" as a function of disturbance does not apply in many ecosystems, especially of those in arid and semi-arid regions (Box 4.1). Rather, sites are characterized by multiple steady states, and stochastic events influence the rate and path of succession. Soil properties also influence the rate and path of succession, so that patterns of species replacement may be altered by management activities which affect soils (e.g., livestock grazing and timber harvesting reduce organic matter and increase bulk density). Furthermore, discontinuous and irreversible transitions in ecosystem structure may occur.

An example illustrates the multiplicity of factors that influence succession (Figure 6 in Archer 1989). In the presence of livestock herbivory, many sites dominated by tallgrass species are replaced by successively shorter and more grazing-resistant herbs. A reduction in grazing intensity may contribute to succession along a similar path as retrogression, especially if fire is maintained within the community. However, continued grazing pressure leads to a reduction in grass cover and, therefore, decreased fire frequency and an increased rate of woody plant establishment. Such grassland-to-woodland transitions have been widely documented during the last 150 years (Archer 1995b; McPherson 1997). On domination of a site by woody plants, reduction in grazing intensity will not initiate succession back to grassland. Establishment of grass plants is reduced by the shade of the dominant woody plants, and grass cover is therefore insufficient to allow fires to spread and kill woody plants. Thus, "recovery" of grassland requires considerable cultural energy (e.g., mechanical or chemical treatments to reduce woody plant abundance and subsequent sowing or planting of grasses). It should be clear that determination of the successional pathway is not merely an academic exercise: imagine, for example, the case where management

---

Box 4.1 Managing with state-and-transition models

The state-and-transition model has direct application to resource management (Westoby *et al.* 1989). Consider, for example, management of livestock in semi-arid ecosystems. Many such systems are characterized by at least two distinct stable states: savanna and woodland (Archer 1990, 1995b; McPherson 1997). Savannas support substantial livestock production whereas woodlands are poorly suited to this activity.

Livestock grazing, in concert with fire suppression and possibly above-average precipitation, may cause conversion of savanna to woodland. Simply reducing grazing pressure is insufficient to cause conversion back to savanna, as incorrectly predicted by the Clementsian model of succession. Rather, the conversion of savanna to woodland is irreversible in the absence of major cultural inputs such as herbicides or mechanical removal of woody plants. Thus, the woodland state is permanent over temporal scales relevant to management.

Of the many consequences of a transition from savanna to woodland, the most apparent is that the management activity responsible for the transition (livestock grazing) cannot be used to reverse the transformation. Further, the woodland is poorly suited to livestock production because herbaceous production is relatively low and woody plants interfere with pastoralism. Thus, managers of savannas should be aware that a common consequence of livestock grazing is reduction of a site's capacity to support continued grazing. One prospective solution is proactive management to remove woody plant seedlings (e.g., with periodic prescribed fires). However, this strategy requires careful planning and short-term loss of livestock forage (via cessation of grazing and consequent combustion of herbaceous plants).

---

objectives include high levels of livestock grazing on sites such as those described by Archer (1989).

At about the same time Westoby *et al.* (1989) published their state-and-transition model, Tilman was delivering an invited lecture based on his research at Cedar Creek Natural History Area (CCNHA). A modified version of his lecture was published shortly thereafter (Tilman 1990). Tilman's "trade offs" approach grew out of his resource-ratio hypothesis of competition (Tilman 1985). Tilman recognized four constraints on the establishment and growth of plants: colonization, which included many of the constraints detailed by Pickett *et al.* (1987); availability of limiting soil resources; availability of light; and sources of death (e.g., herbivores, pathogens). There are six possible two-way trade offs, four possible three-way trade offs, and one possible four-way trade off between these constraints. Tilman systematically eliminated hypotheses via experimentation, and concluded that a three-way trade off between colonization, nutrient competition, and light competition dictates the rate and path of old-field succession at CCNHA. Specifically, colonization (order of arrival of species) and ability to acquire nitrogen in the presence of neighbors determine the successional pathway for grasses. The transition from grassland to oak woodland is best explained by the trade off between nutrient acquisition and light acquisition.

Tilman's approach and conclusions are noteworthy from several viewpoints. The large amounts of time and money invested in his experiments virtually ensure that they will not be replicated elsewhere. Tilman's model has more classes than previous models, reflecting the perception that ecosystem change is a complex process. In addition, the trade off used to describe succession changes over time. Finally, Tilman concludes that nearly all of the constraints he recognized contributed to the pattern of succession at CCNHA. His conclusion about succession at CCNHA has profound implications for resource management (Tilman 1990:14): "Other plant communities will have other constraints, and other successions will be explained by other processes."

THE CONTEMPORARY VIEW

Succession involves the recruitment of a set of species which has a different mortality (or different rate of mortality) than a different set of species. Thus, death of some species, and the concomitant replacement by other species, leads to changes in species composition. The key components of succession are recruitment and mortality; both these factors are affected by species' natural histories, interactions between species and stochastic processes.

Tilman's "trade offs" model reflects the modern consensus that succession is tightly linked with interactions between plants (e.g., "competition" appears in the title of his 1990 paper). Thus, succession is dependent on the responses of individual species to other plants and to the environment. This introduces considerable complexity into discussions of succession. In addition, the stochastic nature of environmental events ensures that succession will be characterized by unpredictability and additional complexity. The power of the state-and-transition model lies in its ability to accommodate site-specific phenomena. However, effective application of the state-and-transition model requires considerable flexibility in management actions (Westoby *et al.* 1989) and a willingness to develop detailed models on a site-specific basis (Joyce 1992). Constraints on applicability and inherent complexity have no doubt contributed to resistance in adoption of the state-and-transition model: although the organismic view of communities has been largely discredited, it continues to appear in various forms (e.g., Wilson and Sober 1989; Wilson 1997). The persistence of this paradigm is a testimony to its simplicity and to the absence of a conceptually simple and readily useful model to replace it.

Despite the limited ability to generalize about succession, several factors often affect the rate and direction of ecosystem change. These include the species composition at the time of disturbance and the type, intensity, frequency, and scale of disturbance events. Unfortunately, many of the generalizations derived from these phenomena are too coarse scaled to be useful for management (e.g., large disturbances benefit wind-dispersed seeds). In addition, interactions between these factors may obscure patterns associated with their main effects.

McIntosh's (1980) article on the history of succession research contains several insightful statements about succession and science. He cites Frank Egler as saying "ecology may not only be more complicated than we think, it may be more complicated than we can think" (McIntosh 1980:53). He captures the pessimism and frustration of ecologists in citing Frank Golley (1977:53): "A simple mechanistic explanation of succession is not possible." Nonetheless, McIntosh concludes that "the search for clarity if not unity in succession has daunted ecologists from the beginning" (McIntosh 1980:53), and he provides some encouragement (p. 54): "the search for satisfying regularity and simplicity is traditional in science, and there is no reason to forgo that search." Thus, we may never generate a simple, generally applicable model for succession, but that is not an adequate reason for abandoning efforts to understand the process. The remainder of this chapter will describe various approaches used to study succession (Figure 4.3).

TOOLS TO STUDY SUCCESSION

Retrospective approaches

Retrospective studies of succession include historical accounts, repeat photography, dendrochronology, and analysis of organic carbon or

Figure 4.3 Dense, "over-stocked" coniferous forest in the Blue Mountains of eastern Oregon. A tradition of fire suppression often leads to densely packed stands with high fuel loads, which are susceptible to catastrophic wildfires or outbreaks of insects. Photo by Stephen DeStefano.

biogenic opal in soil. These approaches cannot be used to test hypotheses about ecosystem dynamics, although they may be useful for describing changes that have occurred on specific sites and for generating hypotheses. Retrospective techniques can typically be used to assess species-level changes in plant distribution only with dominant woody plants. A few studies of vegetation change are, however, exceptional in their fine taxonomic resolution and spatial scale (e.g., Neilson and Wullstein 1983; Neilson 1986; Wondzell and Ludwig 1995). Nonetheless, these efforts are similar to other retrospective studies in that they are correlative and therefore cannot be used to distinguish between the many confounding factors associated with succession. Additional limitations of specific types of retrospective studies are described below.

Historical accounts of ecosystem change (e.g., land survey records, early maps, and notes of early travelers, surveyors, and military scouts) are usually anecdotal and imprecise, and thus do not allow the accurate determination of historical vegetation physiognomy or species composition. In addition, historical accounts are often contradictory and colored by fallacies (Bahre 1991).

Repeat ground photography has a limited and oblique field of view, and historical photographs usually portray anthropogenic manipulation of landscapes. These characteristics seriously limit the usefulness of repeat photography for determining changes in the distribution of species (Bahre 1991). Repeat aerial photography is also constrained by the date of the earliest photographs. In addition, extensive coverage of aerial photographs was not available until after broad-scale ecosystem changes had already occurred.

Dendrochronology is limited to woody plants, usually trees, and is based on correlations between tree age and cross-sectional ring number. Dendrochronological assessments are used to describe the dates of establishment, defoliation, or stem injury of individual woody plants. These assessments are then extrapolated to stands of trees in an attempt to describe periods of recruitment, mortality, rapid growth, or disturbance (Fritts 1976; Johnson and Gutsell 1994). However, if trees were once present but are currently absent, then reconstructions of stand age structure cannot be used to elucidate this important change. Perhaps more importantly, the characteristics of dominant woody plants in many ecosystems are poorly suited for dendrochronological assessment because: (1) current dendrochronological techniques are usually unsuitable for the determination of the stem age of several species, and (2) stem age does not necessarily represent individual plant age, since many species resprout after top removal.

Analyses of stable carbon isotopes have been used to assess vegetation change in grasslands and savannas. Stable isotope analysis relies on differential fractionation of carbon isotopes during photosynthesis. Nearly all woody plants possess the C_3 pathway of photosynthesis, whereas the dominant grasses in subtropical and tropical ecosystems have the C_4 metabolic pathway. These two metabolic pathways ultimately affect the stable carbon isotope ratio ($^{13}C/^{12}C$) of living plant tissue, which is retained and incorporated into soil organic material after plant mortality and decomposition. Therefore, the stable carbon isotope ratio in the soil can be used as an indicator of previous vegetation on a site (Figure 4.4).

The isotopic composition of soil organic carbon does not accurately reflect the past dynamics of C_3 and C_4 vegetation if: (1) the isotopic composition of the surface soil differs from that of the overlying vegetation; (2) soil depth is not an appropriate surrogate for time (if relatively new carbon is transported beneath older soil carbon via soil mixing – e.g., soils may be mixed by burrowing animals, freezing and thawing cycles, or alluvial processes); (3) deep-rooted C_3 plants (e.g., shrubs, trees) deposit soil carbon beneath C_4 plants; or (4) current or former dominant grasses possess the C_3 photosynthetic pathway. Because these conditions often occur within some ecosystems, stable isotope analysis is not appropriate for studying succession within these systems (Dzurec *et al.*

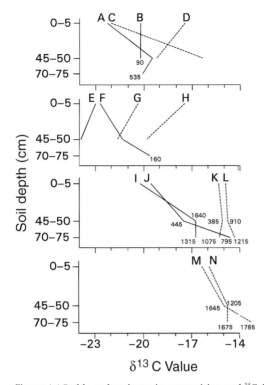

Figure 4.4 Stable carbon isotopic composition and ¹⁴C dates for soil organic matter at different soil depths along a transect through a *Quercus* savanna into a C₄ semi-desert grassland in southern Arizona, United States. Each line represents a single soil core. Soil cores A–D are 500 m above the savanna–grassland ecotone, cores E–H are 150 m above the ecotone, I–L are at the savanna–grassland ecotone, and cores M and N are 150 m below the ecotone in the grassland. Solid lines are cores collected beneath trees and dashed lines are cores collected beneath grasses >5 m from a tree canopy. Values adjacent to lines are ¹⁴C dates (in years) at different depths; all depths without dates had postmodern signatures. Reproduced with permission from McClaran and McPherson (1995).

1985; McClaran and McPherson 1995). However, analysis of stable carbon isotopes is useful for identifying past shifts in boundaries between systems dominated by plants with different photosynthetic pathways (e.g., subtropical forest/grassland boundaries).

Vegetation changes may be inferred by assessing biogenic opal (i.e., plant microfossils or opal "phytoliths") in soils (e.g., Kalisz and Stone 1984), and the technique is conceptually similar to stable isotope analysis. Grasses produce more biogenic opal than woody plants, and the opal from grasses is morphologically distinct from the opal of woody plants (Witty and Knox 1964; Kalisz and Stone 1984). Biogenic opal is comprised of silica dioxide, which is very resistant to decomposition. Thus, the abundance and type of opal in the soil can be used to indicate previous vegetation on a site. Biogenic opal can be used to distinguish between some members of the grass family, and has, therefore, been used to study conversions from perennial to annual grasslands (e.g., Bartolome *et al.* 1986). The limitations of biogenic opal analysis are similar to those of stable isotope analysis: both techniques rely on chemical or morphological differences between plant taxa (especially grasses and woody plants) and make similar assumptions about deposition in the soil.

The concurrent use of several different retrospective techniques may facilitate the appropriate interpretation of past changes in ecosystems. However, different retrospective techniques may generate conflicting interpretations of the same phenomena, as illustrated by the following example (McPherson and Weltzin 2000):

> Reports of past changes in the oak savanna/semidesert grassland boundary are varied. Paleoecological data suggest that oak savannas have shifted upslope in concert with warmer and drier conditions since the Pleistocene. This interpretation is consistent with upslope movement of most woody species in the last 40,000 years, as determined by paleoecological research (Betancourt *et al.* 1990). In contrast, research based on stable carbon isotope technology and radiocarbon dating indicated that oaks at the savanna/grassland boundary had encroached into former grasslands within the last 1,500 years, which implied that oak savannas had shifted downslope into semidesert grasslands (McPherson *et al.* 1993; McClaran and McPherson 1995). The latter finding matches Leopold's (1924) interpretation of downslope movement of oaks, based on observations of progressively smaller trees from the savanna into the grassland. On a more contemporary temporal scale, use of repeat ground photography led Hastings and Turner (1965) to conclude that the oak savanna/semidesert grassland boundary moved upslope during the last century. Finally, Bahre (1991) examined surveyor's records, repeat ground photography, and repeat aerial photography, and concluded that the distribution of oak savannas had been stable since the 1870s. Thus, boundaries between oak savannas and adjacent semidesert grasslands have been variously reported as shifting upslope, remaining static, or shifting downslope. Although these differences may be attributable in part to variation in temporal and spatial scales, they are largely the result of different interpretations.

This example indicates that the disadvantages associated with retrospective techniques cannot be overcome simply with the use of multiple methods.

There are two primary, overarching limitations to using retrospective approaches to study succession. First, it is virtually impossible to reconstruct accurately the events and conditions that contributed to past changes, even at well-studied localities. Second, even if this were possible, conditions responsible for historical or prehistoric species distributions are unlikely to be repeated in the future. Earth is entering an unprecedented era in terms of atmospheric gas concentrations, climatic conditions, land use, and land cover; thus, even a complete understanding of past climates and assemblages of organisms will not allow the confident prediction of future changes. This situation is exacerbated by species-specific response patterns that are often not linear or predictable, even within life forms (Tilman and Wedin 1991; Archer 1993). Finally, results of retrospective investigations do not elucidate mechanisms of ecosystem changes (*sensu* Simberloff 1983; Campbell *et al.* 1991) because confounding between various factors precludes identification of mechanisms and introduces the potential for spurious correlations.

Consider the relatively recent large-scale changes in vegetation physiognomy that have occurred in former grasslands and savannas throughout the world. Dramatic transitions from grasslands and savannas to closed-canopy woodlands have captivated the scientific community, but the mechanisms underlying the changes remain unknown after more than three decades of detailed investigation (Archer 1989). For example, increased woody plant abundance in most grasslands and savannas has been attributed to changes in atmospheric or climatic conditions, reduced fire frequency, increased livestock grazing, or combinations of these factors (as reviewed by Archer 1994). Differing opinions about the causes of vegetation change have contributed to acrimonious debate. For example, Bahre (1991:105), in a critique of work conducted by Hastings and Turner (1965), concluded that "probably more time has been spent on massaging the climatic change hypothesis than on any other factor of vegetation change, and yet it remains the least convincing." Such debate hardly seems beneficial for scientific advancement or appropriate management, yet it is a natural product of retrospective approaches.

Regardless of scientific progress toward consensus on the mechanisms of past ecosystem change, elucidation of these mechanisms would provide little or no predictive power to current and future management of natural ecosystems. Events that may have contributed to past changes in ecosystem structure (e.g., cattle grazing, decreased fire frequency, specific timing of precipitation) may fail to produce similar responses today because of other, more profound changes in the physical and biological environments over the last century. For example, ecosystems now experience increased atmospheric concentrations of greenhouse gases (e.g., carbon dioxide, methane, nitrous oxides), increased abundance of woody perennial plants and introduced plants, and decreased abundance of some plant and animal species. Finally, there are no historical analogs for the conditions which are now widespread. For example, a return to the fire regimes which characterized prehistoric ecosystems is unlikely to occur without major cultural inputs to extant ecosystems. Even if we could reproduce prehistoric fire regimes, the decision to do so should be based on clearly defined, site-specific goals and objectives, rather than on the misguided hope that restoration of disturbance regimes will necessarily restore prehistoric ecosystems. Our knowledge of the past should guide contemporary management, not constrain it (Figure 4.5).

The descriptive nature of retrospective approaches, coupled with the complex site-specific interactions underlying ecosystem change, render these approaches unsuitable for the determination of the mechanisms of ecosystem change. Retrospective approaches are constrained by fundamental conceptual and philosophical limitations, and they are hampered by various technical weaknesses. Several of the technical obstacles associated with specific retrospective techniques have been removed by significant technological advances, and we expect continued progress in this area of research; however, the more important conceptual and philosophical limitations can be overcome only by traveling back in time. Thus, although it is widely acknowledged that understanding mechanisms of ecosystem change is central to the interpretation and prediction of species and ecosystem responses to disturbance or climate change, there is no evidence to suggest that such mechanistic understanding can be achieved with retrospective approaches.

Monitoring

Monitoring involves direct observations of changes in ecosystem structure over time. As such, managers usually monitor in an attempt to discover and document changes in ecosystem structure as they occur. This information serves as a primary basis for evaluating and possibly altering management strategies. Thus, monitoring is an important component of adaptive management.

Many attributes can be monitored, including soil, life forms, and species. However, monitoring programs usually focus on species composition. This approach is similar in most respects to the retrospective

Figure 4.5 Second-growth forest of Douglas-fir (*Pseudotsuga menziesii*) and western hemlock (*Tsuga heterophylla*) following timber harvest in the coastal mountain range of western Oregon. Species composition at the time of disturbance, land-use patterns, and the nature of perturbations over time contribute to postdisturbance vegetation patterns. Photo by Stephen DeStefano.

approaches discussed above, with two notable exceptions: (1) spatial and temporal scales generally are finer and (2) taxonomic resolution usually is higher. These two attributes engender confidence that changes in ecosystem structure result from events that precede the changes. For example, a reduction in livestock density that precedes a shift in species composition may be interpreted as causal. The confidence associated with this conclusion typically increases with decreased time between sampling intervals, increased site specificity, and increased taxonomic resolution. However, this interpretation must be tempered with the knowledge that this approach is correlative and phenomenological, even with infinitely short sampling intervals, complete knowledge of changes on a particular site, and perfect taxonomic resolution. Considerable caution is warranted before changes in management are invoked as causal. Shifts in species composition or other measures of ecosystem structure that are attributed to management may have resulted from subtle and unperceived environmental events or other confounding factors. Management is not an experiment, regardless of the degree to which the management is "adaptive:" treatments are not assigned at random to experimental units, experimental units are rarely homogeneous or replicated, and managers frequently change many factors simultaneously. These characteristics preclude confident determination of causality.

Although changes in species composition may not be confidently attributed to management, observing these changes is an important component of effective management. Several metrics can be monitored, as described in Chapter 3. Ordination can then be used to describe changes in species composition over time (*sensu* Austin 1977). The direction and distance that quadrats "move" in ordination space over time may reflect successional patterns (Figure 4.6).

Comparative studies

Succession has frequently been inferred from studies that compare sites with different elapsed times after disturbance (i.e., chronosequences, or "space-for-time" substitutions). The taxonomic resolution of these comparative studies is usually superior to the resolution associated with retrospective approaches, and is similar to that obtained with monitoring. Comparative studies are similar to retrospective approaches in several respects; in particular, they are useful for generating hypotheses and for describing changes that have occurred on specific sites, and they cannot be used to test hypotheses about succession. This approach assumes that all sites under study have identical histories of disturbance, biotic influence, and environmental conditions (Luken 1990); this assumption is rarely valid, which makes this approach susceptible to serious criticism (e.g., Miles 1979). Nonetheless, chronosequences may be the best approach for describing successional sequences that predate Anglo settlement, and they have been widely used to suggest patterns of vegetation change in terrestrial ecosystems with long-lived organisms (Burrows 1990).

Comparative studies in the mesic central United States have suggested that old-field succession can be described on the basis of changes in models of community organization. Specifically, early-successional plant communities are characterized by a geometric model; as succession

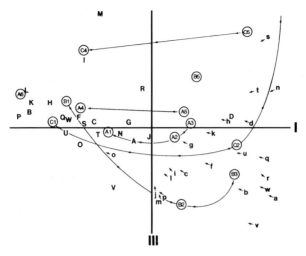

Figure 4.6 Reciprocal averaging ordination of desert shrub vegetation in Guadalupe Mountains, Texas, United States. Transects sampled in 1972 are indicated by lowercase letters, and transects sampled in 1980 are indicated by uppercase letters. Arrows attached to the 1972 transects point in the direction of their 1980 counterparts. Species class/form class variables are circled, with the following legend: A, decreases with livestock grazing; B, increases with livestock grazing; C, invades with livestock grazing; 1–3 = all forage available; 4–6 = forage partially available; 1, 4 = little or no hedging; 2, 5 = moderately hedged; and 3, 6 = severely hedged. Incomplete arrows attached to species class/form class sequences result from scaling of figure. Reproduced with permission from Wester and Wright (1987).

proceeds, the plant community is described by a log-normal curve which becomes successively steeper over time (Figure 4.7). Thus, the models of community structure described in Chapter 3 appear to be useful not only for characterizing communities in space, but also for describing how communities change over time.

Comparative research is hampered by weak inference (Platt 1964): concluding that observed differences in species composition result directly from elapsed time ignores the potential impact of many other factors. As such, the results of comparative research do not provide reliable mechanistic explanations. Differences in postdisturbance climatic conditions may have a greater impact on species composition than time since disturbance, and inherent differences between sites may either exacerbate or obscure the effect of elapsed time. Climate and site factors undoubtedly interact with time since disturbance, and therefore cannot

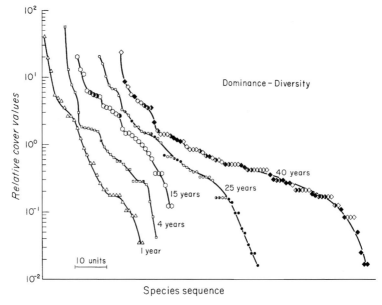

Figure 4.7 Dominance–diversity curves for old fields of five different ages of abandonment in southern Illinois, United States. Unfilled symbols are herbs, half-filled symbols are shrubs, and closed symbols are trees. Reproduced with permission from Bazzaz (1975).

be ruled out as candidate explanations for postdisturbance differences in species composition.

The three approaches described thus far – retrospection, monitoring, and comparison – are descriptive techniques. Although descriptive studies are necessary and important for describing ecosystem structure and identifying hypotheses about ecosystem dynamics, strict reliance on descriptive research severely constrains the ability of ecology to solve managerial problems. Descriptive studies rely on comparison of patterns and subsequent invocation of mechanisms. This process is not reliable when many hypotheses make similar predictions about observed patterns; because nature is complex, it is almost always possible to develop alternative hypotheses. In addition, the poor predictive power of ecology (Peters 1991) indicates that our knowledge of ecosystem function is severely limited (Stanley 1995). Unjustified reliance on descriptive research and the inability to understand ecosystem function are among the most important obstacles that prevent ecology from making significant progress toward solving environmental problems and from being a predictive science. Many ecologists (e.g., Hairston 1989; Keddy 1989; Gurevitch and Collins 1994) have concluded that field-based manipulative experiments represent a logical approach for future research. Unfortunately, there have been few ecological experiments focused on succession.

The inability of descriptive research to serve as a basis for hypothesis testing does not negate the importance of this approach in research and management. In fact, a carefully planned and executed monitoring protocol is a fundamental aspect of effective, site-specific management. Objective and quantifiable metrics of ecosystem structure and function provide the foundation for the assessment of the effectiveness of management in meeting site-specific objectives. Nonetheless, it must be recognized that even the most precise and accurate monitoring efforts cannot be used for testing hypotheses.

Experiments

Experiments are necessary to determine the mechanisms of ecosystem change (see Chapter 1). A common prediction in ecology with direct application to management is that different management activities (i.e., land uses) affect succession in a specified manner. Ideally, a manipulation is performed to evaluate this prediction. Unfortunately, few ecological experiments have focused on succession.

IMPLICATIONS FOR WILDLIFE POPULATIONS

The processes, mechanisms, and results of vegetation succession have important implications for the abundance, distribution, and structure of wildlife populations and animal communities. This statement is intuitive to the point of being obvious, yet few observational studies have explicitly demonstrated how succession can influence wildlife populations, presumably because of the long-term nature of succession. Even fewer experimental studies have been conducted because of logistical issues and constraints associated with time and money.

Newton's (1991, 1993) long-term research on the European sparrowhawk (*Accipiter nisus*) is a rare example that demonstrates explicit links between changes in vegetation structure and distribution and demography of an avian species. The European sparrowhawk is a woodland raptor that eats mainly small birds and is distributed throughout Great Britain and much of Europe. For about 20 years, Newton banded and observed breeding sparrowhawks, particularly females, and monitored their nest sites, productivity, and movements. Newton's research has shown that sparrowhawk nests located in young woods with small densely growing trees had the highest occupancy rates and reproductive success, but, as woodlands matured and trees became larger and more widely spaced, both occupancy and success declined. Furthermore, removal experiments confirmed the presence of nonbreeding individuals in the population; these individuals attempted to nest when sites in younger woods became available, but would otherwise remain as nonbreeders despite the presence of vacant sites in older forest stands. Thus, the quality of the habitat (in this case, young woods with small, densely growing trees) was important to breeding success. As succession progressed, the quality of the habitat for breeding sparrowhawks declined.

Although there may be few other examples of the effects of succession on wildlife distribution and demography as explicit as Newton's sparrowhawks, the dynamic forces that shape the composition and structure of plant communities obviously have important impacts on animal communities (Box 4.2). These forces and processes, if understood by wildlife managers, can be manipulated to the benefit of some wildlife populations. In addition, there are some additional key points that should be considered by those interested in managing wildlife populations.

Box 4.2 Early successional stages as wildlife habitat

In most parts of the temperate region of the world, a forest covering a broad region (i.e., a forest covering hundreds of km², undisturbed by humans) would provide a wide array of cover types for a large number of species. Although many environmental variables, such as soil type, hydrology, and elevation, and many natural disturbances, such as fire or wind, are involved in determining the distribution and abundance of species, it is often the *structure* of the vegetation that ecologists tend to measure to gain some understanding of community ecology.

Historically, wildlife managers recognized that many animals, particularly some sought-after game species such as grouse, quail, and deer, were more abundant where *edge* habitat was prevalent. Edge habitat describes those areas or ecotones where two or more cover types meet, such as mature forest, second-growth forest, and fields. For many species, this is ideal habitat, and in fact the "intermediate disturbance hypothesis" predicts that diversity will be greatest where disturbance is intermediate. For example, a forest with a mixture of mature stands of trees, second-growth trees, regenerating clearcuts, and either anthropogenic or natural openings will have greater species diversity than either a uniform old-growth forest (minimum disturbance) or a large clearcut (maximum disturbance). Throughout the middle of the last century, wildlife managers often manipulated vegetation to maximize edge.

In recent decades, however, the loss of older forests – those woodlands characterized by large trees, dense canopy cover, large standing and downed dead wood (i.e., snags and logs) – and the concern for species dependent on those older forests – such as spotted owls, northern goshawks, and red tree voles – has led to public pressure to reduce human-caused disturbances such as logging. This has led to widespread concern for habitat fragmentation, especially in forested ecosystems (Harris 1990).

Even more recently, biologists have become concerned with the loss of early successional communities (Askins 2001). As disturbed areas such as old farm fields and young stands of trees such as aspen disappear from the landscape through both natural succession and lack of periodic disturbances such as fire, tree harvest, or mowing, these early successional communities are disappearing from the landscape, especially in eastern North America (Trani *et al.* 2001). Along with them go an assemblage of many early-successional stage or disturbance-dependent species (Hunter *et al.* 2001; Litvaitis 2001). Biologists are now recommending ways to perpetuate and maintain these early successional communities on the landscape (Thompson and DeGraaf 2001).

First, the term "habitat" is often loosely used by wildlife biologists. It has been defined as the sum total of all the environmental components used by a species for its life history. Thus, references to cover types such as pine habitat or oak habitat used by sparrowhawks, or the desert, montane, or forested habitat of bobcats (*Felis rufus*), are too vague and narrow, and do not conform to the habitat concept. Hall *et al.* (1997:175) offered the following definition:

> We therefore define "habitat" as the resources and conditions present in an area that produce occupancy – including survival and reproduction – by a given organism. Habitat is organism-specific; it relates the presence of a species, population, or individual (animal or plant) to an area's physical and biological characteristics. Habitat implies more than vegetation or vegetation structure; it is the sum of the specific resources that are needed by organisms.

Thus, the concept of "habitat" is much more than just plant cover. For some species, habitat can be quite complex – e.g., habitat for migratory

Figure 4.8 Habitat for bobcat, a wide-ranging species found in many different environments, includes any place where resources and conditions allow the species to survive and reproduce. Photo by Stephen DeStefano.

species includes areas needed not only for breeding, but for migration and wintering as well (Figure 4.8).

Second, the concept of habitat "quality" is increasingly interesting to wildlife biologists. An evaluation of habitat quality is often based on the demographic performance of the species of interest; if individuals show optimum reproductive output and high survivorship, then that habitat is thought to be of high quality (DeStefano *et al.* 1995). High density of individuals does not necessarily mean that those individuals are in high-quality habitat or will show high breeding success (van Horne 1983; Vickery *et al.* 1992a). However, beyond generalizations such as high abundance and availability of food resources, well-distributed nest or den sites, or adequate safety from predators or inclement weather, biologists have not been very successful in determining the exact characteristics of a habitat that make it high quality (although see Vickery *et al.* 1992b). Nonetheless, the concept of habitat "quality" or "fitness" (DeStefano *et al.* 1995) is important to management and conservation and deserves further study.

Third, although a complete definition of the term habitat implies more than vegetation or vegetation structure, it is usually vegetation that we try to manipulate and manage for wildlife populations. Extant vegetation is a product of historical events, plant propagules, and ecological

Figure 4.9 Red tree voles occupy the canopies of coniferous forests in the Pacific Northwest of the United States. Fragmentation of the forest canopy can affect the distribution, abundance, and viability of red tree vole populations. Photo by Stephen DeStefano.

interactions (although the relative roles of these forces are frequently unknown). For example, fire regime, intensity of livestock grazing, and silvicultural systems shape the resulting community – this is the "legacy" idea to which forest ecologists sometimes refer.

Fourth, stochastic events, such as wildfires, windstorms, or floods, can greatly alter the rate and path of succession and can change wildlife habitat. This realization is especially important for small populations, including species threatened with extinction, which are often the major targets of management efforts. Extinction is a deterministic process, often punctuated by a stochastic event, which frequently results from habitat loss or fragmentation (Figure 4.9).

Fifth, Newton's studies of site fidelity among sparrowhawks showed how tenacious individual birds can be in their "loyalty" to a home site, and how beneficial this tenacity can be in terms of reproductive output and longevity. Many wildlife species display an incredible capacity to return to or stay at a site even after alteration, and there can be a delayed response after even severe changes in vegetation structure. Examples include a pair of northern goshawks returning to a nest site that was mature forest the previous year, but is now on the edge of a clearcut, and sage grouse (*Centrocercus urophasianus*) returning to a lekking ground that has been paved over since their last spring ritual.

It is important to remember, from a wildlife management perspective, that the outcome of vegetation succession and the subsequent effects on local animal populations will vary in time and space; that is, vegetation change and consequent alterations in animal communities are spatially and temporally specific, and these processes take place in a fluctuating environment (Morrison *et al.* 1998). We are not suggesting that managers are destined to "reinvent the wheel" at each specific time and location; rather, managers must acknowledge and appreciate modern succession theory, establish clear and realistic goals, and incorporate site-specific characteristics such as land-use history, past management techniques (e.g., prescribed burning, timber harvest patterns, soil disturbance), and presence of nonnative species. Few management plans involve long-term monitoring of the vegetation and responses of targeted wildlife populations. Efforts to alter the successional patterns of local plant communities should not be attempted without a well-designed and realistic monitoring program for plant and animal populations (Morrison *et al.* 1998; Thompson *et al.* 1998).

SUMMARY

Understanding and predicting changes in ecosystems over time are necessary components of effective management. These tasks are not easily accomplished, however: multiple, interacting factors affect succession, and the resulting complexity is daunting. Science may never develop a comprehensive model of succession that is relevant to site-specific management. Such a model seems fundamental to effective management, and the absence of a model necessitates decision-making in the absence of complete scientific information. Thus, natural resource management is an art as well as a science; if management of natural resources is done with a conscience, it is among the most difficult of human endeavors. Gaps between science and management impose fundamental constraints on effective management, and closing the gap between these enterprises will enhance both of them.

Figure 4.10 Alteration of successional patterns by a keystone species: foraging and dam-building by beavers alter the landscape in ways that are beneficial to many species of wildlife but which are sometimes in conflict with humans. Photo by Stephen DeStefano.

Chapter 5 will identify sources of the divide between ecology and natural resource management, and establish a foundation for narrowing the divide (Figure 4.10).

5

Closing the gap between science and management

In his popular book "A Brief History of Time," the internationally renowned physicist Stephen W. Hawking (1988) forwards a view that is popular among physical scientists: ". . . the eventual goal of science is to provide a single theory that describes the whole universe." This Theory of Everything would obviously benefit the management of human activities, but it is not clear what contribution ecology might make to this theory. In ecology, as in most sciences, increased levels of organization are characterized by decreased scientific precision and increased complexity of arrangement. For example, boundary recognition and spatial arrangement of individual organisms are relatively simple and straightforward; in contrast, ecosystems typically have complex spatial arrangements and boundaries that are difficult to discern. Chris Beckett (1990) responds to Hawking's statement with a relatively pessimistic outlook for science: "And then we get to the levels in which we actually live out our lives: our relationships, our aspirations, politics, . . . personal choices, moral dilemmas . . . And at this level (the most important one, after all, from a human perspective) science has no precise answers, no complete descriptions at all." Clearly, our expectations of science as a source of knowledge must be tempered with an understanding that the natural world will never be completely described. Because prediction typically represents an even greater challenge than description, it is tempting to abandon ecology as a source of practical managerial information.

The inability to apply ecological information to environmental problems is vexing and frustrating to scientists who generate knowledge and to managers who attempt to apply that knowledge. As a result, the divide between the science of ecology and management of natural resources is vast, despite increasingly frequent pleas for ecologists to focus on managerial problems (e.g., Keddy 1989; Lubchenco *et al.* 1991; Kessler *et al.* 1992; Sharitz *et al.* 1992; Underwood 1995) (Figure 5.1).

Figure 5.1 Using an increment borer to count rings and estimate the age of trees. Photo by Stephen DeStefano.

The gap between science and management is evident at several levels, and can be easily recognized within the discipline of ecology. The problems addressed by applied ecologists are frequently viewed by theoretical ecologists as site specific or poorly grounded in ecological theory. However, theoretical ecologists have increasingly distanced themselves from applied disciplines by: ignoring the historical and contemporary role of humans in ecological processes; downplaying the importance of social, political, and economic processes on ecological processes; and discounting or ignoring the literature associated with applied disciplines (e.g., forestry, range science, wildlife management). Thus, theoretical ecologists have "discovered" various phenomena years or decades after their acceptance in the management community, and have tended to conduct research that is perceived to be irrelevant by natural resource managers. For example, the concept of carrying capacity was described by range scientists in the 1890s, developed by wildlife biologists in the 1920s, and finally "discovered" by ecologists in the 1950s (Young 1998). Of course, theoretical ecologists and applied ecologists are simply points at the ends of a continuum: most individual ecologists willingly identify numerous individuals on either end of the continuum relative to themselves.

We believe that both ecology and management can benefit significantly from the other enterprise, and that the rift between them is particularly detrimental to the effective management of natural resources. Thus, this chapter will describe some of the reasons for the rift between science and management, and offer approaches to help close the gap between the two endeavors.

The ultimate test of ecology is whether ecologists can say anything useful about the natural world. According to one prominent plant ecologist: "we can develop all the elegant models we wish, live distinguished academic careers, publish numerous well-cited papers, and so on, but the ultimate test of the value of our work is whether we really can make predictions about the real world" (Keddy 1989:157). According to this view, applications are the primary tool with which to evaluate the progress of ecology, but can ecological theory be applied to environmental problems?

Two goals of community ecology that are relevant to management are the development of assembly rules and the development of response rules (Keddy 1989). The objective of assembly rules is to predict the abundance of organisms based on knowledge of the species pool and the environment (Diamond 1975). The objective of response rules is to predict future community composition based on knowledge of the current species composition, the total species pool, and a specific disturbance or land use. The total species pool includes species currently present on a site and nearby species that are capable of dispersing into the site following a disturbance or change in land use.

It should be evident that these goals of community ecology are implicit goals of natural resource management. Managers must create or maintain ecosystems that are capable of producing a variety of products and services. To do so, they must be able to predict the effects of various land uses on the abundance of species, including species in the region that are not currently present on a specific site.

Both assembly rules and response rules require considerable knowledge of key life-history traits in the species pool. They also require knowledge of the way in which species interact with various types of disturbances. As such, they rely on a combination of description, comparison, and experimentation (Keddy 1989:156–7). Description is used to delineate the species pool, define initial states of systems, and to describe naturally occurring states that result from various land uses (cf. state-and-transition model, Chapter 4). Comparison of the attributes of species generates the necessary ecological information on species in the pool. Finally, experimentation is needed to determine how species respond to different kinds of land uses.

Figure 5.2 Setting prescribed fire with a drip-torch. Photo by
Guy R. McPherson.

Several questions illustrate the potential role of ecology in natural
resource management. What size of reserve is sufficient to protect different groups of species? Which kinds of species will be the first to disappear due to alterations in atmospheric gas concentrations, climate,
fire regimes, or levels of livestock grazing? What are the states and transitions associated with specific sites? What are the implications of these
states and transitions for management? Answering these questions is
necessary for effective management, and also would contribute to the
development of ecological theory. Most importantly, these questions ex-
emplify the ultimate test of ecology: they allow ecologists to say some-
thing useful about the natural world (Figure 5.2).

EVALUATING PROGRESS

Evaluating progress represents a significant challenge to most scientific
disciplines. Incorporating criteria about applying scientific information (i.e., management) adds additional complexity. Can the "progress
of science" be evaluated objectively? What about progress in applying
science?

Keddy (1989:157) suggests that we start to evaluate progress by
dividing ideas into two classes: hypotheses and concepts. Hypotheses are
falsifiable statements that represent candidate explanations for patterns
observed in nature (Chapter 1). Concepts are not falsifiable, although

they are part of every person's thinking. Concepts provide a framework which helps to organize hypotheses, and may lead people to new creative insights. However, scientific data can not be used to resolve issues that are based on different concepts.

The value that we place on different kinds of questions and different kinds of approaches depends on the relative emphasis that we place on hypotheses and concepts. Concepts are useful if we see science as an activity which expands the horizons of human experience. In this case, we can be satisfied if we increase our understanding of nature. However, evaluating understanding or passing on increased understanding may be difficult or impossible. Alternatively, we may seek to describe and predict the processes that underlie patterns observed in nature. Hypotheses are a fundamental component of such a predictive ecology.

To date, ecology has placed considerable emphasis on concepts and little emphasis on hypothesis testing (Peters 1980; Keddy 1989). One result is that ecologists "have become modern scholastics interminably discussing questions which cannot be solved or tested scientifically" (Peters 1980). Such insoluble questions can not contribute to the solution of environmental problems.

Judging the value of different research goals and methods also requires consideration of how scientific progress actually occurs. There is no consensus on this point, and numerous views have been described according to their position along a continuum (Keddy 1989):

1. Science primarily involves the patient collection of facts. There may also be the belief that someone will eventually make sense of them through induction. This view values data for their own sake, and is synonymous with natural history.

2. Data are important for falsifying hypotheses, and original hypotheses drive scientific progress. Data are collected to falsify hypotheses, and their collection is guided by the question being asked. This view is modeled after Popper (1959).

3. Data are collected to solve minor technical problems, but there is a larger context or paradigm shared by scientists. Data are collected to clarify aspects of the paradigm, but not to challenge it.

4. Science is primarily political. Individuals of perceived high status dictate the prevailing world view, and data are collected to substantiate this world view. Contrary data are rarely collected, are not accepted, and can not be published.

5.   Science is part of the entertainment industry, and the objective of scientific papers is to tell entertaining stories to a well-educated audience. Short papers which pose a clear question and provide an answer are labeled as "naive" or as "least publishable units."

Data are paramount at one end of this continuum, whereas they are collected only to amplify belief systems at the other end. Although different perspectives are appropriate for different situations, extreme views appear to hold less promise for advancing science than more moderate views. In some situations, strict adherence to inappropriate views may hinder scientific progress.

THE RELEVANCE OF ECOLOGY TO NATURAL RESOURCE MANAGEMENT

As indicated in Chapter 1, this book does not provide explicit recommendations for resource managers for two important reasons. First, management decisions must be temporally, spatially, and objective specific. Thus, this book should not be used for site-specific management decisions; rather, management decisions should be couched within this temporally and spatially broad discussion and should be made by managers most familiar with individual systems (*sensu* McPherson and Weltzin 2000). Second, specific management activities, although presumably based on scientific knowledge, are conducted within the context of relevant social, economic, and political issues. These specific issues and concerns are beyond the scope of this book, which is instead focused on scientific knowledge.

As discussed by McPherson and Weltzin (2000), the realm of science represents a substantial reservoir of relatively untapped information available to resource managers. As such, managers in need of scientific information are encouraged to use existing data, work closely with the scientific community, and communicate the need for specific information. Further, it is critical that resource managers understand how scientific knowledge is obtained: effective managers should be familiar with scientific principles. For example, not all scientific information will enable managers to predict accurately the response of an ecosystem to a specific disturbance or manipulation. Some research findings actually present untested hypotheses rather than observed responses to well-controlled experimental manipulations. The results of such research should be interpreted judiciously (Chapter 1).

Figure 5.3 Identifying grassland plants along a transect. Photo by Kiana Koenen.

The development of theory is intended to be as general as possible. In fact, the discovery of universal scientific laws (e.g., evolution by natural selection) represents a powerful goal for science. The application of ecological theory is necessarily site and objective specific; in addition, management must be conducted within the context of relevant social, economic, and political issues. Resolving the paradox between the generality of theory development and the specificity of theory application represents the crux of the problem for applied ecology (Figure 5.3).

The paradox between the generality of theory development and the specificity of theory application is exemplified in virtually every issue of scientific journals. Few managers read the ecological literature because the research reported therein does not appear to be relevant to natural resource management. We offer three familiar examples. First, heated debate has developed with respect to the relative importance of the size and number of reserves required to meet conservation goals. The single-large or several-small (SLOSS) debate has filled hundreds of pages in leading ecological journals. The volume of literature dedicated to this topic implies that it must be important, yet an overwhelming majority of managers will never contribute to the design of a conservation reserve. In the rare cases when new conservation reserves are established, we suspect that the SLOSS literature – which is characterized by debate about simple ecological models which trivialize the natural history of important species – is ignored in the process of their design and establishment. Second, the relationship between species diversity and ecosystem function has received much attention from ecologists. Consensus has not been reached on the importance of species diversity to ecosystem function, much less on a cause for a relationship. In fact, there is no reason to expect a single, general relationship between species diversity and ecosystem properties (e.g., productivity, rates of nutrient cycling) because the relative contributions of species to ecosystem properties are strongly influenced by the environment (Cardinale *et al.* 2000). In other words, the environmental context within which species establish, grow, and interact has greater influence on ecosystem function than the absolute number of species. A surprising amount of energy and expense has been spent to discover that the identity and characteristics of species exert considerable control over ecosystem function: that species "matter" has been known at least since Aristotle's time. Third, journals are similarly replete with case studies of specific species or ecosystems. Paramount to publication of research in the "best" journals are the ecologist's ability to couch the research question within the context of contemporary ecological theory and the cleverness of the experimental design used to address the question. Considerably less important is the creative application of research results beyond the system under scrutiny (and sometimes even within this system). It is small wonder that overworked managers dedicate little time to the study of ecological literature.

Given the overwhelming volume of ecological literature, it is easy to understand why managers rarely consult literature that is not based on local research. Few managers are interested in the SLOSS debate (similarly arcane examples are plentiful), and many consider the results of case studies to be site specific and of no particular importance to management in local systems. This reflects a recurring theme in ecology: the quest for general principles (which is valued in ecology, as in other sciences) necessarily involves the study of details. As a result, scientists frequently end up making natural history observations instead of contributing to ecological theory and, in the process of illuminating details, all hope for generality seems to vanish. Of course, this problem is not restricted to ecology. Different people appreciate different degrees of generality, so that one person's conceptual richness is another person's trivial detail. Anyone who has been subjected to the wedding videos of friends can understand the roles of familiarity and perception on the appreciation for detail.

Different levels of specificity are necessary for different management questions and ecological scales. Ultimately, ecological theory may consist of a series of nested conceptual models (Keddy 1989). Specific models geared toward the management of individual systems will use site-specific information and precise taxonomic resolution. These models will be particularly useful for the development of site-specific management strategies. Presumably, they would be nested within more general conceptual models which would incorporate relationships among state variables and functional groups of organisms. The latter models would be useful for establishing and assessing landscape-level policies.

CONSTRAINTS ON APPLYING ECOLOGY

Managers must recognize that there are many questions that science can not answer. For example, science can not dictate which elements of nature merit conservation. Determination of what to conserve depends on human values (Lawton 1997; McPherson 2001b). An exponential human growth rate precludes preservation of all genetic diversity, species, and ecosystems: humans use a disproportionate percentage of Earth's resources, which contributes to the loss of biological diversity at all levels (Vitousek *et al.* 1986). Ecologists can provide objective information about objects or processes under consideration for conservation, but the ultimate choice about *whether* or *how much* to conserve is beyond the realm of ecology. This debate is centered on the extent to which natural resources should be used to support human activities. Once society has made a decision about what to conserve, ecology can provide the tools for evaluating the success of conservation efforts. These tools will have maximum utility when the objectives are clearly stated and when they can be quantified (Chapter 1).

Ecology can serve as a framework for addressing many questions relevant to management, but there are constraints on this process. For example, the speed of scientific inquiry rarely matches the urgency of environmental problems. In addition, the complex nature of most environmental problems precludes simple solutions that can be easily applied to many sites. Finally, scientific findings are subject to revision or reinterpretation: managers become frustrated and may abandon ecology as a source of information when it appears that ecologists change their minds about the relevant facts.

Frustration notwithstanding, resource managers require reliable scientific information to manage ecological communities and processes effectively. The volume of available data on these topics is overwhelming, and individual managers must identify and extract the relevant information in order to address management issues. Additional factors contribute to the difficult dilemma that managers face as they attempt to incorporate scientific knowledge into management decisions: (1) much of the available information is contradictory or inconsistent, and (2) many scientists still attempt to provide mechanistic explanations about ecosystem function based on descriptive research. This latter tendency has trapped scientists into making predictions about things they cannot predict (Peters 1991; Underwood 1995). Adherence to scientific principles, including hypothesis testing, will improve communications between resource managers and scientists while increasing the credibility of both groups.

The terminology of scale

Confusion and misunderstanding between ecologists and managers result at least partially from different perceptions of scale and associated differences in terminology (Allen and Hoekstra 1992). The selection of labels for levels of organization is a human activity: there are no absolute levels of organization, independent of the observer. Further, the selection of a level depends on the phenomena of interest. For example, ecologists and managers could use guilds or trophic levels instead of communities as the basis for study and communication, depending on the specific phenomena under observation. Confusion results from the inherent subjectivity associated with identifying and communicating about levels of organization.

Contrary to typical interpretations, the labels that are commonly used to describe ecological phenomena are not necessarily size ordered. For example, a forest *community* may contain a single *organism* (a rotting tree) that encompasses several *ecosystems*. Similarly, a single *organ* of a ruminant *organism* (the rumen) may contain an *ecosystem*. To "size order" these attributes (*sensu* Figure 5.4) is to define the scale of interest; this process represents a definition, but it may or may not accurately represent nature. This would not be problematic, except for the considerable difficulty in linking levels of organization (Allen and Hoekstra 1992). Common phenomena must be used to link different levels (e.g., cycling of a specific element, or growth). Terminology of phenomena must also be consistent – e.g., individual competition at the level of individuals at temporal scales of years is not equivalent to competition at the level of populations over evolutionary time. Strict adherence to labels, in the absence of appropriate context, generates difficulty in communication and, therefore, confusion.

The "tower" model of organization (Figure 5.4) provides a familiar example with which to illustrate this problem. This model clearly does

Figure 5.4 Eight levels of organization in ecology arranged in a straight tower.

not represent a size-ordered view of the world. It is intended to provide generality, yet it actually accounts for relatively little flow of energy or materials through ecosystems (Allen and Hoekstra 1990, 1992). We recognize that the use of traditional terminology will continue (e.g., labels associated with the tower model), and we believe that this use is appropriate if it is recognized that most labels are arbitrary, subjective categories that are not scale defined. In fact, use of traditional terminology is preferred over the development of new terms in the absence of new concepts (McIntosh 1980) (Figure 5.5).

Ecology and management within a socio-political context

There are some indications that ecology is primarily a "social" activity, which is consistent with Beckett's (1990) view about all scientific activities. For example, ecology is characterized by "invisible colleges" of colleagues that influence the development and resulting application of ecological theory (McIntosh 1980). An extreme interpretation of this view suggests that ecological theory is largely a function of sociology, an arena in which science is severely constrained. Ecologists influence the outcome of scientific investigations by selecting the level of study (e.g., organisms, populations, ecosystems) and the phenomena of interest (e.g., growth, energy flow). Ideally, these decisions are made with an aim toward the generation of reliable knowledge via hypothesis testing (*sensu* Chapter 1).

Figure 5.5 Forest canopy cover can be estimated with an ocular tube or "moosehorn" (Garrison 1949; Cook *et al.* 1995). Photo by Stephen DeStefano.

The phenomena that are observed and described depend on the level of organization selected by the observer (Allen *et al.* 1984). For example, a physiological ecologist who chooses to focus on leaf-level phenomena will probably advance our understanding of population dynamics very slowly or not at all. Similarly, an ecologist who studies trends in populations over space or time will probably fail to uncover mechanistic explanations for changes in populations at lower or higher levels of organization. Because few individuals have the intellectual capacity or energy to understand numerous levels of organization, important ecological knowledge probably lies undiscovered by the relevant investigators. Inability to link ecological subdisciplines hinders the development of ecology as either a predictive or explanatory science. Similarly, natural resource managers do not have sufficient time to read the ecological literature and link together information from various sources. The resulting failure to integrate information across several levels of organization undoubtedly constrains the effective application of ecological information.

Ecologists exert considerable influence on the patterns and processes that are discovered and described because investigators select the domain of interest. Such selection is usually done with little or no input from the management community. For example, ecologists have been captivated with the study of competition for at least a century, and this fascination has come at the expense of studies on other processes. Several hypotheses have been offered to explain the focus on competition (Keddy 1989): (1) scientists are influenced by their culture, and ecology has developed rapidly in the United States; (2) competition is obviously interesting to people, especially in contrast to other ecological processes; (3) gender bias in the male-dominated field of ecology has favored studies of aggression rather than, for example, mutualism; (4) ecological research is highly atypical of organisms occupying the earth, with a taxonomic bias in favor of vertebrates; (5) competition within the scientific community has selected for aggressively competitive individuals; and (6) elitism within the ranks of ecology has allowed relatively few individuals to set the agenda for the entire discipline.

Ecologists also select the grain and extent of studies, and these attributes determine the limits on the spatial and temporal scale of observable phenomena. Coarse-grained studies (e.g., studies that rely on satellite imagery) can not detect fine-grained phenomena (e.g., dietary requirements of a specific bird species), and fine-grained studies are ineffective for the determination of large or long-term processes (e.g., successional pathways).

Obstacles to communication may also interfere with the progress of ecology and its application. The inherent subjectivity of peer review makes the process vulnerable to bias and inconsistency. In some disciplines, the perceived status of authors and consistency of the results with orthodox views may be more important than the quality of the research in determining whether research is published (Mahoney 1976; Peters and Ceci 1982; Keddy 1989). Although the process of peer review has not been studied in ecology, there is no reason to believe that ecologists are less susceptible to bias than other scientists. A partial solution to this problem would be double-blind reviews of research proposals and manuscripts, which are common in some disciplines but rare in ecology and management.

Despite the presence of these constraints on the development of scientific knowledge, we do not hold a postmodernist philosophy about ecology, and we are not suggesting that ecology is characterized by scientific relativism (*sensu* Dennis 1996). Ecology operates within the sphere of a single physical universe characterized by facts, patterns, and processes that are known or knowable by all observers. Although the *pace* of discovery is influenced by the socio-political and cultural environment, we do not question the *existence* of these discoveries or the associated facts. In this way, science is distinct from art: whereas a significant artistic contribution could come only from a specific artist, a scientific contribution will be discovered (if not by a particular scientist, then by another at a later time). However, we believe that some controversies can be resolved by recognizing the relative importance of socio-political or cultural factors in the debate.

APPLYING ECOLOGY

We have reached a critical juncture in the management of natural resources. Managers are drowning in a sea of information, but much of the information is not relevant to management. It is incumbent on managers to determine which elements of the expansive ecological literature are relevant to specific management objectives, and to extrapolate concepts generated elsewhere into activities at a particular site. Ecologists have begun to make a commitment to applied environmental research, and they need guidance and support from the management community. Managers and ecologists stand to benefit from the other, but they are struggling to break down old barriers and to "connect" with each other. Fortunately, most of the impediments to ecologically based management of natural resources are related to the psychology and sociology of investigators and managers – and these obstacles can be overcome with patience, persistence, and enhanced communication. It seems clear that progress will not occur by collecting yet another data set consistent with established dogma, or by continuing rhetorical debates. Careful, informed thought, rather than tradition and habit, should serve as a basis for natural resource management and should guide the selection of research questions, systems, and conceptual approaches.

For maximum effectiveness, resource managers must understand how scientific knowledge is obtained. For example, not all information generated by scientists will enable managers to predict accurately the response of a plant community to a specific disturbance or manipulation. Some research findings present untested hypotheses rather than documented responses of different ecosystems to different disturbances. In fact, most research journals encourage authors to describe potential mechanisms for observed patterns. Managers and policy-makers routinely confuse these tentative, untested hypotheses for tested, documented phenomena and use the former as a basis for decisions. Such reliance on untested hypotheses may be necessitated by the absence of reliable knowledge derived from rigorous tests; however, managers should recognize the limitations of these hypotheses. When asked to make predictions, the usual response of scientists is to proceed blithely; if they perceive the problem inherent in making predictions based on tentative hypotheses, they equivocate. Neither blithely proceeding nor equivocating provides useful information for solving management problems.

Fortunately, managers can contribute to scientific inquiry, and therefore bridge the gap between ecology and management, via several specific means. These include posing tractable questions, helping design ecological experiments, seeking management implications from published research, understanding the difference between hypotheses and predictions, understanding weak inference, assessing experimental techniques and research methods, and facilitating insightful research experiments on lands within their jurisdiction.

Ecologists, too, can take steps to link science and management. For example, scientists can conduct research within the context of local managerial problems, and thus use local ecosystems to examine how plant or animal populations might respond to experimental manipulations. Experiments are often conducted under a certain set of conditions and the results are published, after which the scientist moves on to other projects: the generality of experimental findings is rarely evaluated. In fact, a cynic might conclude that the primary products of scientific research are controversy, confusion, and publications – in other words, the aim of most research is to generate discussion or produce papers in scientific journals, rather than answer specific questions or further the understanding of complex natural systems (Hobbs 1998). In contrast to this cynical view, researchers can work with local land managers to assess the generality of their work by attempting to predict the response of plant or animal populations under conditions that differ from their original experiment. This effort could have significant benefits for both science and management. In addition to the generation of reliable knowledge via experimentation and routine assessment of experimental research, ecologists should continue the development of monitoring protocols to evaluate the success of management actions. Scientists should encourage managers to use monitoring protocols that are based on measurable, clear objectives (e.g., to identify changes in species composition over time) (Wicklum and Davies 1995; Lélé and Norgaard 1996; McPherson 1997). Finally, scientists

Figure 5.6 Trained dogs help a biologist locate quail. Photo by Stephen DeStefano.

can facilitate management by focusing on questions that address important management issues within the context of a mechanistic program of research, synthesizing relevant research from their research and that conducted by other scientists, supplying information in outlets accessible to managers, and responding to requests for information and advice in a timely and thoughtful manner.

The world needs general, predictive ecological theory for the conservation and appropriate use of natural resources, but the development of such a predictive ecology represents a monumental undertaking. Further, natural resource management is among the most difficult of human endeavors. Managers hold the key to conservation biology, including the retention of biodiversity, maintenance of ecosystem function, and production of services and products for human use. Thus, ecologists and managers can be satisfied that their efforts are important to the effective management of earth's ecosystems. They should be inspired by the pressure and the challenge imposed by current and future generations (Figure 5.6).

References

Aarssen, L. W. (1989). Competitive ability and species coexistence: a 'plant's-eye' view. *Oikos* **56**:386–401.

Aarssen, L. W. (1992). Causes and consequences of variation in competitive ability in plant communities. *Journal of Vegetation Science* **3**:165–74.

Aarssen, L. W., and Epp, G. A. (1990). Neighbour manipulations in natural vegetation: a review. *Journal of Vegetation Science* **1**:13–30.

Aarssen, L. W., and Turkington, R. (1985). Competitive relations among species from pastures of different ages. *Canadian Journal of Botany* **63**:2319–25.

Abbott, L. B., and McPherson, G. R. (1999). Nonnative grasses in southern Arizona: historical and contemporary perspectives. In *A Century of Parks in Southern Arizona: Proceedings of the Second Conference on Research and Resource Management in Southern Arizona National Park Areas,* eds. L. Benson and B. Gebow, pp. 3–6. Tucson, AZ: National Park Service.

Adamec, R. E. (1976). The interaction of hunger and preying in the domestic cat. *Behavioral Biology* **18**:263–72.

Alatalo, R. V. (1981). Problems in the measurement of evenness in ecology. *Oikos* **37**:199–204.

Allen, T. F. H., and Hoekstra, T. W. (1990). The confusion between scale-defined levels and conventional levels of organization in ecology. *Journal of Vegetation Science* **1**:5–12.

Allen, T. F. H., and Hoekstra, T. W. (1992). *Toward a Unified Ecology.* New York: Columbia University Press.

Allen, T. F. H., O'Neill, R. V., and Hoekstra, T. W. (1984). Interlevel relations in ecological research and management: some working principles from hierarchy theory. USDA Forest Service Rocky Mountain Research Station General Technical Report RM-110, Fort Collins, CO.

Alpert, P. (1995). Incarnating ecosystem management. *Conservation Biology* **9**:952–5.

Anderson, D. R., and Burnham, K. P. (1992). Data-based selection of an appropriate biological model: the key to modern data analysis. In *Wildlife 2001: Populations,* eds. D. R. McCullough and R. H. Barrett, pp. 16–30. New York: Elsevier Applied Science.

Anderson, D. R., Burnham, K. P., Gould, W. R., and Cherry, S. (2001). Concerns about finding effects that are actually spurious. *Wildlife Society Bulletin* **29**:311–16.

Anderson, D. R., Burnham, K. P., and Thompson, W. L. (2000). Null hypothesis testing: problems, prevalence, and an alternative. *Journal of Wildlife Management* **64**:912–23.

Andrewartha, H. G. (1961). *Introduction to the Study of Animal Populations.* Chicago: University of Chicago Press.

Archer, S. (1989). Have southern Texas savannas been converted to woodlands in recent history? *American Naturalist* **134**:545–61.

Archer, S. (1990). Development and stability of grass/woody mosaics in a subtropical savanna parkland, Texas, USA. *Journal of Biogeography* **17**:453–62.

Archer, S. (1993). Vegetation dynamics in changing environments. *Rangelands Journal* **15**:104–16.

Archer, S. (1994). Woody plant encroachment into southwestern grasslands and savannas: rates, patterns, and proximate causes. In *Ecological Implications of Livestock Herbivory in the West*, eds. M. Vavra, W. Laycock and R. Pieper, pp. 13–68. Denver: Society for Range Management.

Archer, S. (1995a). Harry Stobbs Memorial Lecture, (1993): herbivore mediation of grass–woody plant interactions. *Tropical Grasslands* **29**:218–35.

Archer, S. (1995b). Tree-grass dynamics in a *Prosopis*-thornscrub savanna parkland: reconstructing the past and predicting the future. *Ecoscience* **2**:83–99.

Arrhenius, O. (1921). Species and area. *Journal of Ecology* **9**:95–9.

Arthur, W., and Mitchell, P. (1989). A revised scheme for the classification of population interactions. *Oikos* **56**:141–3.

Askins, R. A. (2001). Sustaining biological diversity in early successional communities: the challenge of managing unpopular habitats. *Wildlife Society Bulletin* **29**:407–12.

Aspinall, D., and Milthorpe, F. L. (1959). An analysis of competition between barley and white persicaria. I. The effects on growth. *Annals of Applied Biology* **47**:156–72.

Austin, M. P. (1976). Non-linear species response models in ordination. *Vegetatio* **33**:33–41.

Austin, M. P. (1977). Use of ordination and other multivariate descriptive methods to study succession. *Vegetatio* **35**:165–75.

Austin, M. P. (1986). The theoretical basis of vegetation science. *Trends in Ecology and Evolution* **1**:161–4.

Austin, M. P., and Gaywood, M. J. (1994). Current problems of environmental gradients and species response curves in relation to continuum theory. *Journal of Vegetation Science* **5**:473–82.

Austin, M. P., Nicholls, A. O., Doherty, M. D., and Meyers, J. A. (1994). Determining species response functions to an environmental gradient by means of β-function. *Journal of Vegetation Science* **5**:215–28.

Bahre, C. J. (1991). *A Legacy of Change: Historic Human Impact on Vegetation in the Arizona Borderlands.* Tucson, AZ: University of Arizona Press.

Banks, P. B., Dickman, C. R., and Newsome, A. E. (1998). Ecological costs of feral predator control: foxes and rabbits. *Journal of Wildlife Management* **62**:766–72.

Bartlett, M. S. (1947). The use of transformations. *Biometrics* **3**:39–52.

Bartolome, J. W., Klukkert, S. E., and Barry, W. J. (1986). Opal phytoliths as evidence for displacement of native Californian grassland. *Madroño* **33**:217–22.

Bazzaz, F. A. (1975). Plant species diversity in old-field successional ecosystems. *Ecology* **56**:485–8.

Beckett, C. (1990). The great chain of being. *New Scientist* **125**:60–1.

Beissinger, S. R., and Westphal, M. I. (1998). On the use of demographic models of population viability in endangered species management. *Journal of Wildlife Management* **62**:821–41.

Bender, E. A., Case, T. J., and Gilpin, M. E. (1984). Perturbation experiments in community ecology: theory and practice. *Ecology* **65**:1–13.

Betancourt, J. L., Van Devender, T. R., and Martin, P. S. (1990). *Packrat Middens: The Last 40,000 Years of Biotic Change*. Tucson, AZ: University of Arizona Press.

Bibby, C. J., Burgess, N. D., and Hill, D. A. (1992). *Bird Census Techniques*. London: Academic Press.

Blackburn, T. M., and Gaston, K. J. (1996). Abundance–body size relationships: the area you census tells you more. *Oikos* **75**:303–9.

Boertje, R. D., Valkenburg, P., and McNay, M. E. (1996). Increases in moose, caribou, and wolves following wolf control in Alaska. *Journal of Wildlife Management* **60**:474–89.

Bonham, C. D. (1989). *Measurements for Terrestrial Vegetation*. New York: Wiley.

Boutin, S., Krebs, C. J., Sinclair, A. R. *et al.* (1995). Population changes of the vertebrate community during a snowshoe hare cycle in Canada's boreal forest. *Oikos* **74**:69–80.

Brand, C. J., Keith, L. B., and Fischer, C. A. (1976). Lynx responses to changing snowshoe hare densities in Alberta. *Journal of Wildlife Management* **40**:416–28.

Brewer, R., and McCann, M. T. (1985). Spacing in acorn woodpeckers. *Ecology* **66**:307–8.

Briske, D. D., and Richards, J. H. (1995). Plant responses to defoliation: a physiologic, morphologic and demographic evaluation. In *Wildland Plants: Physiological Ecology and Developmental Morphology*, eds. D. J. Bedunah and R. E. Sosebee, pp. 635–710. Denver: Society for Range Management.

Bronstein, J. L. (1994). Conditional outcomes in mutualistic interactions. *Trends in Ecology and Evolution* **9**:214–17.

Brown, D. (1954). *Methods of Surveying and Measuring Vegetation*. Bucks, UK: Commonwealth Agricultural.

Brown, J. R., and MacLeod, N. D. (1996). Integrating ecology into natural resource management policy. *Environmental Management* **30**:289–96.

Brunner, R. D., and Clark, T. W. (1997). A practice-based approach to ecosystem management. *Conservation Biology* **11**:48–58.

Buckland, S. T., Anderson, D. R., Burnham, K. P., and Laake, J. L. (1993). *Distance Sampling: Estimating Abundance of Biological Populations*. London: Chapman and Hall.

Bulla, L. (1994). An index of evenness and its associated diversity measure. *Oikos* **70**:167–71.

Burgess, J. W. (1983). Reply to a comment by R. L. Mumme *et al. Ecology* **64**:1307–8.

Burgess, J. W., Roulston, D., and Shaw, E. (1982). Territorial aggregation: an ecological spacing strategy in acorn woodpeckers. *Ecology* **63**:575–8.

Burnham, K. P., and Anderson, D. R. (1992). Data-based selection of an appropriate biological model: the key to modern data analysis. In *Wildlife 2001: Populations*, eds. D. R. McCullough and R. H. Barrett, pp. 16–30. New York: Elsevier Applied Science.

Burnham, K. P., and Anderson, D. R. (1998). *Model Selection and Inference: A Practical Information Theoretic Approach*. New York: Springer-Verlag.

Burrows, C. J. (1990). *Processes of Vegetation Change*. London: Unwin Hyman.

Burton, P. J., Balisky, A. C., Coward, L. P., Cumming, S. G., and Kneeshaw, D. D. (1992). The value of managing for biodiversity. *Forestry Chronicle* **68**:225–37.

Camargo, J. A. (1993). Must dominance increase with the number of subordinate species in competitive interactions? *Journal of Theoretical Biology* **161**:537–42.

Campbell, B. D., Grime, J. P., Mackey, J. M. L., and Jalili, A. (1991). The quest for a mechanistic understanding of resource competition in plant communities: the role of experiments. *Functional Ecology* **5**:241–53.

Campbell, D. J. (1995). Detecting regular spacing in patchy environments and estimating its density using nearest-neighbour graphical analysis. *Oecologia* **102**:133–7.

Canfield, R. H. (1941). Application of the line interception method in sampling range vegetation. *Journal of Forestry* **39**:388–94.

Cardinale, B. J., Nelson, K., and Palmer, M. A. 2000. Linking species diversity to the functioning of ecosystems: on the importance of environmental context. *Oikos* **91**:175–83.

Carleton, T. J., Stitt, R. H., and Nieppola, J. (1996). Constrained indicator species analysis (COINSPAN): an extension of TWINSPAN. *Journal of Vegetation Science* **7**:125–30.

Carpenter, S. R. (1996). Microcosm experiments have limited relevance for community and ecosytem ecology. *Ecology* **77**:677–80.

Carpenter, S. R., Chisholm, S. W., Krebs, C. J., Schindler, D. W., and Wright, R. F. (1995). Ecosystem experiments. *Science* **269**:324–7.

Casado, M. A., Ramírez-Sanz, L., Castro, I., de Miguel, J. M., and de Pablo, C. L. (1997). An objective method for partitioning dendrograms based on entropy parameters. *Plant Ecology* **131**:193–7.

Castillo, D. (2001). Population estimates and behavioral analyses of managed cat colonies located in Miami-Dade County, Florida, Parks. M. S. Thesis, Miami: Florida International University.

Causton, D. R. (1988). *An Introduction to Vegetation Analysis: Principle, Practice, and Interpretation.* London: Unwin Hyman.

Cherry, S. (1996). A comparison of confidence interval methods for habitat use-availability studies. *Journal of Wildlife Management* **60**:653–8.

Cherry, S. (1998). Statistical tests in publications of The Wildlife Society. *Wildlife Society Bulletin* **26**:947–53.

Clements, F. E. (1916). *Plant Succession: An Analysis of the Development of Vegetation.* Washington, DC: Carnegie Institute of Washington Publication 242.

Clements, F. E., Weaver, J. E., and Hanson, H. C. (1929). *Plant Competition: An Analysis of Community Functions.* Washington, DC: Carnegie Institute of Washington Publication 398.

Cole, J., Lovett, G., and Findlay, S. (eds.) (1991). *Comparative Analyses of Ecosystems.* Berlin: Springer-Verlag.

Coleman, J. S., and Temple, S. A. (1993). Rural residents' free-ranging domestic cats: a survey. *Wildlife Society Bulletin* **21**:381–90.

Colinvaux, P. (1986). *Ecology.* New York: Wiley.

Connell, J. H. (1961). The influence of interspecific competition and other factors on the distribution of the barnacle *Chthamalus stellatus. Ecology* **42**:710–23.

Connell, J. H. (1990). Apparent vs. "real" competition in plants. In *Perspectives on Plant Competition,* eds. J. B. Grace and D. Tilman, pp. 9–26. New York: Academic Press.

Connell, J. H., and Slatyer, R. O. (1977). Mechanisms of succession in natural communities and their role in community stability and organization. *American Naturalist* **111**:1119–44.

Conner, M. M., Jaeger, M. M., Weller, T. J., and McCullough, D. R. (1998). Effect of coyote removal on sheep depredation in northern California. *Journal of Wildlife Management* **62**:690–9.

Connor, E. F., and Simberloff, D. (1979). The assembly of species communities: chance or competition? *Ecology* **60**:1132–40.

Cook, J. G., Stutzman, T. W., Bowers, C. W., Brenner, K. A., and Irwin. L. L. (1995). Spherical densiometers produce biased estimates of forest canopy cover. *Wildlife Society Bulletin* **23**:711–17.

Cortwright, S. A. (1988). Intraguild predation and competition: an analysis of net growth shifts in larval amphibian prey. *Canadian Journal of Zoology* **66**:1813–21.

Cottam, G., and Curtis, J. T. (1956). The use of distance measures in phytosociological sampling. *Ecology* **37**:451–60.

Cowles, H. C. (1899). The ecological relations of the vegetation on the sand dunes of Lake Michigan. Parts 1, 2, 3, 4. *Botanical Gazette* **27**:95–117, 167–202, 281–308, 361–91.

Crooks, K. R., and Soulé, M. E. (1999). Mesopredator release and avifaunal extinctions in a fragmented system. *Nature* **400**:563–6.

Dale, M. R. T. (1984). The continuity of upslope and downslope boundaries of species in a zoned community. *Oikos* **42**:92–6.

Daubenmire, R. (1968). *Plant Communities*. New York: Harper and Row.

Daw, S. K., DeStefano, S., and Steidl, R. J. (1998). Does survey method bias the description of goshawk nest-site structure? *Journal of Wildlife Management* **62**:1379–84.

Dennis, B. (1996). Discussion: should ecologists become Bayesians? *Ecological Applications* **6**:1095–103.

DeStefano, S. (1998). Determining the status of northern goshawks in the West: is our conceptual model correct? *Journal of Raptor Research* **32**:342–8.

DeStefano, S., and McCloskey, J. (1997). Does vegetation structure limit the distribution of northern goshawks in the Oregon Coast Ranges? *Journal of Raptor Research* **31**:34–9.

DeStefano, S., Thrailkill, J. A., Swindle, K. A., Miller, G. S., Woodbridge, B., and Meslow, E. C. (1995). Analysis of habitat quality and relative survival using capture–recapture data. In *Integrating People and Wildlife for a Sustainable Future,* eds. J. A. Bissonette and P. R. Krausman, pp. 466–9. Proceedings of the First International Wildlife Management Congress. Bethesda, MD: The Wildlife Society.

Diamond, J. M. (1975). Assembly of species communities. In *Ecology and Evolution of Communities*, eds. M. L. Cody and J. M. Diamond, pp. 342–4. Cambridge, MA: Belknap Press, Harvard University Press.

Diamond, J. M. (1986). Overview: laboratory experiments, field experiments and natural experiments. In *Community Ecology*, eds. J. M. Diamond and T. J. Case, pp. 3–22. New York: Harper and Row.

Diamond, J. M., and Gilpin, M. (1982). Examination of the 'null' model of Connor and Simberloff for species co-occurrences on islands. *Oecologia* **52**:64–74.

Drake, J. A. (1990). Communities as assembled structures: do rules govern pattern? *Trends in Ecology and Evolution* **5**:159–64.

Drury, W. H., and Nisbet, I. C. (1973). Succession. *Journal of the Arnold Arboretum* **54**:331–68.

Dueser, R. D., Porter, J. H., and Dooley, J. L., Jr. (1989). Direct tests for competition in North American rodent communities: synthesis and prognosis. In *Patterns in the Structure of Mammalian Communities*, eds. D. W. Morris, Z. Abramsky, B. J. Fox and M. R. Willig, pp. 105–25. Special Publication No. 28. Lubbock, TX: Texas Tech University Press.

Dyksterhuis, E. J. (1949). Condition and management of rangeland based on quantitative ecology. *Journal of Range Management* **2**:104–15.

Dzurec, R. S., Boutton, T. W., Caldwell, M. M., and Smith, B. N. (1985). Carbon isotope ratios of soil organic matter and their use in assessing community composition changes in Curlew Valley, Utah. *Oecologia* **66**:17–24.

Eckblad, J. W. (1991). How many samples should be taken? *BioScience* **41**:346–8.

Egler, F. E. (1954). Vegetation science concepts. I. Initial floristic composition – a factor in old-field vegetation development. *Vegetatio* **4**:412–17.

Evans, E. W., and Seastedt, T. R. (1995). In *Wildland Plants: Physiological Ecology and Developmental Morphology*, eds. D. J. Bedunah and R. E. Sosebee, pp. 580–634. Denver: Society for Range Management.

Forsman, E. D., S. DeStefano, S., Raphael, M. G., and Gutiérrez, R. J. (eds.) (1993). Demography of the northern spotted owl. *Studies in Avian Biology* **17**:1–122.

Fritts, H. C. (1976). *Tree Rings and Climate.* New York: Academic Press.

Garrison, G. A. (1949). Uses and modifications for the "moosehorn" crown closure estimator. *Journal of Forestry* **47**:733–5.

Gauch, H. G., Jr. (1982). *Multivariate Analysis in Community Ecology.* Cambridge: Cambridge University Press.

Gauch, H. G., Jr., and Stone, E. L. (1979). Vegetation and soil pattern in a mesophytic forest at Ithaca, New York. *American Midland Naturalist* **102**:332–45.

Gauch, H. G., Jr., and Whittaker, R. H. (1981). Hierarchical classification of community data. *Journal of Ecology* **69**:537–57.

Gause, G. F. (1930). Studies on the ecology of the Orthoptera. *Ecology* **11**:307–25.

Gavin, T. A. (1989). What's wrong with the questions we ask in wildlife research? *Wildlife Society Bulletin* **17**:345–50.

Gavin, T. A. (1991). Why ask "why": the importance of evolutionary biology in wildlife science. *Journal of Wildlife Management* **55**:760–6.

Gilpin, M. E., and Diamond, J. M. (1982). Factors contributing to the non-randomness in species co-occurrences on islands. *Oecologia* **52**:75–84.

Gleason, H. A. (1917). The structure and development of the plant association. *Bulletin of the Torrey Botanical Club* **43**:463–81.

Gleason, H. A. (1926). The individualistic concept of the plant association. *Bulletin of the Torrey Botanical Club* **53**:7–26.

Golley, F. B. (1965). Structure and function of an old-field broomsedge community. *Ecological Monographs* **35**:113–37.

Golley, F. B. (1977). *Ecological Succession.* Stroudsburg, PA: Dowden, Hutchinson and Ross.

Goodall, D. W. (1952). Some considerations of the use of point quadrats in the analysis of vegetation. *Australian Journal of Scientific Research, Series B* **5**:1–41.

Goodall, D. W. (1953). Point-quadrat methods for the analysis of vegetation. *Australian Journal of Botany* **1**:457–61.

Grace, J. B. (1995). In search of the Holy Grail: explanations for the coexistence of plant species. *Trends in Ecology and Evolution* **10**:263–4.

Grace, J. B., and Tilman, D. (eds.) (1990). *Perspectives on Plant Competition.* New York: Academic Press.

Grant, P. R., and Abbott, I. (1980). Interspecific competition, island biogeography and null hypotheses. *Evolution* **34**:332–41.

Grieg-Smith, P. (1983). *Quantitative Plant Ecology*, 3rd edn. Berkeley, CA: University of California Press.

Grime, J. P. (1979). *Plant Strategies and Vegetation Processes.* Chichester: Wiley.

Grimm, V. (1994). Mathematical models and understanding in ecology. *Ecological Modelling* **75/76**:641–51.

Grisebach, A. (1838). Ueber den Einfluss des Climas auf die Begranzung der naturlich Floren. *Linnaea* **12**:159–200.

Grubb, P. J. (1977). The maintenance of species-richness in plant communities: the importance of the regeneration niche. *Biological Review* **52**:107–45.

Grumbine, R. E. (1994). What is ecosystem management? *Conservation Biology* **8**:27–38.

Guo, Q., and Rundel, P. W. (1997). Measuring dominance and diversity in ecological communities: choosing the right variables. *Journal of Vegetation Science* **8**:405–8.

Gurevitch, J., and Collins, S. L. (1994). Experimental manipulation of natural plant communities. *Trends in Ecology and Evolution* **9**:94–8.

Gustafson, M. P. (1993). Intraguild predation among larval plethodontid salamanders: a field experiment in artificial stream pools. *Oecologia* **96**:271–5.

Hairston, N. G. Sr. (1989). *Ecological Experiments: Purpose, Design, and Execution.* Cambridge: Cambridge University Press.

Hall, C. A. S. (1988). An assessment of several of the historically most influential theoretical models used in ecology and the data provided in their support. *Ecological Modelling* **43**:5–31.

Hall, C. A. S. (1991). An idiosyncratic assessment of the role of mathematical models in environmental sciences. *Environment International* **17**:507–17.

Hall, L. S., Krausman, P. R., and Morrison, M. L. (1997). The habitat concept and a plea for standard terminology. *Wildlife Society Bulletin* **25**:173–82.

Harlow, L. L., Mulaik, S. A., and Steiger, J. H. (eds.). (1997). *What If There Were No Significance Tests?* Mahwah, NJ: Lawrence Erlbaum Associates.

Harper, J. L. (1977). *Population Biology of Plants.* New York: Academic Press.

Harper, J. L. (1982). After description. In *The Plant Community as a Working Mechanism*, ed. E. I. Newman, pp. 11–25. Oxford: Blackwell.

Harris, L. D. (1990). *The Fragmented Forest: Island Biogeography Theory and the Preservation of Biotic Diversity.* Chicago: University of Chicago Press.

Hastings, J. R., and Turner, R. M. (1965). *The Changing Mile.* Tucson, AZ: University of Arizona Press.

Hawking, S. W. (1988). *A Brief History of Time: From the Big Bang to Black Holes.* New York: Bantam Books.

Hayek, L. -A. C., and Buzas, M. A. (1997). *Surveying Natural Populations.* New York: Columbia University Press.

Heitschmidt, R. K., and Stuth, J. W. (eds.) (1991). *Grazing Management: An Ecological Perspective.* Portland, OR: Timber Press.

Hill, M. O. (1973). Diversity and evenness: a unifying notation and its consequences. *Ecology* **54**:427–32.

Hill, M. O. (1979). TWINSPAN: a FORTRAN program for arranging multivariate data in an ordered two-way table by classification of the individuals and attributes. Ithaca, NY: Ecology and Systematics, Cornell University.

Hill, M. O. (1997). An evenness statistic based on the abundance-weighted variance of species proportions. *Oikos* **79**:413–16.

Hill, M. O., and Gauch, H. G., Jr. (1980). Detrended correspondence analysis: an improved ordination technique. *Vegetatio* **42**:47–58.

Hill, M. O., Bunce, R. G. H., and Shaw, M. W. (1975). Indicator species analysis, a divisive polythetic method of classification, and its application to a survey of native pinewoods in Scotland. *Journal of Ecology* **63**:597–613.

Hobbs, R. J. (1998). Managing ecological systems and processes. In *Ecological Scale: Theory and Applications*, eds. D. L. Peterson and V. T. Parker, pp. 459–84. New York: Columbia University Press.

Hogeweg, P. (1976). Iterative character weighing in numerical taxonomy. *Computers in Biology and Medicine* **6**:199–211.

Holl, K. D. (1998). Do bird perching structures elevate seed rain and seedling establishment in abandoned tropical pasture? *Restoration Ecology* **6**:253–61.

Hotelling, H. (1933). Analysis of a complex of statistical variables into principal components. *Journal of Educational Psychology* **24**:417–41, 498–520.

Hunter, M. L., Jr. (1989). Aardvarks and arcadia: two principles of wildlife research. *Wildlife Society Bulletin* **17**:350–1.

Hunter, W. C., Buehler, D. A., Canterbury, R. A., Confer, J. L., and Hamel, P. B. (2001). Conservation of disturbance-dependent birds in eastern North America. *Wildlife Society Bulletin* **29**:440–55.

Hurlbert, S. H. (1971). The non-concept of species diversity: a critique and alternative parameters. *Ecology* 52:577–86.

Hurlbert, S. H. (1984). Pseudoreplication and the design of field experiments. *Ecological Monographs* 54:187–211.

Hutchinson, G. E. (1959). Homage to Santa Rosalia, or why are there so many kinds of animals? *American Naturalist* 93:145–59.

Hutchinson, G. E. (1961). The paradox of the plankton. *American Naturalist* 95:137–45.

Jackson, D. A., and Somers, K. M. (1991). Putting things in order: the ups and downs of detrended correspondence analysis. *American Naturalist* 137:704–12.

Janzen, D. H. (1986). *Guanacaste National Park: Tropical Ecological and Cultural Restoration.* San José, Costa Rica: Editorial Universidad Estatal a Distancia.

Joern, A. (1986). Experimental study of avian predation on coexisting grasshopper populations (Orthoptera: Acrididae) in a sandhills grassland. *Oikos* 46:243–9.

Johnson, D. H. (1999). The insignificance of statistical significance testing. *Journal of Wildlife Management* 63:763–72.

Johnson, E. A., and Gutsell, S. L. (1994). Fire frequency models, methods and interpretations. *Advances in Ecological Research* 25:239–87.

Jongman, R. H. G., ter Braak, C. J. F., and van Tongeren, O. F. R. (1987). *Data Analysis in Community and Landscape Ecology.* Wageningen, The Netherlands: Pudoc.

Joyce, L. A. (1992). The life cycle of the range condition concept. *Journal of Range Management* 46:132–8.

Kalisz, P. J., and Stone, E. L. (1984). The longleaf pine islands of the Ocala National Forest, Florida: a soil study. *Ecology* 65:1743–54.

Keddy, P. A. (1989). *Competition.* New York: Chapman and Hall.

Keddy, P. A. (1990). Competitive hierarchies and centrifugal organization in plant communities. In *Perspectives on Plant Competition,* eds. J. B. Grace and D. Tilman, pp. 265–90. New York: Academic Press.

Keddy, P. A. (1992). Assembly and response rules: two goals for predictive community ecology. *Journal of Vegetation Science* 3:157–64.

Keddy, P. A., and MacLellan, P. (1990). Centrifugal organization in forests. *Oikos* 59:75–84.

Keiter, R. B. (1995). Greater Yellowstone: managing a charismatic ecosystem. *Natural Resources and Environmental Issues* 3:75–85.

Keith, L. B. (1963). *Wildlife's Ten-year Cycle.* Madison, WI: University of Wisconsin Press.

Keith, L. B. (1983). Role of food in hare population cycles. *Oikos* 40:385–95.

Keith, L. B., Cary, J. R., Rongstad, O. J., and Brittingham, M. C. (1984). Demography and ecology of a declining snowshoe hare population. *Wildlife Monograph* 90:1–43.

Kenkel, N. C., Juhász-Nagy, P., and Podani, J. (1989). On sampling procedures in population and community ecology. *Vegetatio* 83:195–207.

Kennedy, P. L. (1997). The northern goshawk (*Accipiter gentilis atricapillus*): is there evidence of a population decline? *Journal of Raptor Research* 31:95–106.

Kent, M., and Coker, P. (1992). *Vegetation Description and Analysis: A Practical Approach.* Boca Raton, FL: CRC Press.

Kershaw, K. A., and Looney, J. H. H. (1985). *Quantitative and Dynamic Plant Ecology.* London: Edward Arnold.

Kessler, W. B., Salwasser, H., Cartwright, W. C., Jr., and Caplan, J. A. (1992). New perspectives for sustainable natural resources management. *Ecological Applications* 2:221–5.

Krebs, C. J. (1972). *Ecology: The Experimental Analysis of Distribution and Abundance.* New York: Harper and Row.

Krebs, C. J. (1989). *Ecological Methodology.* New York: Harper and Row.

Krebs, C. J., Boutin, S., Boonstra, R. *et al.* (1995). Impact of food and predation on the snowshoe hare cycle. *Science* **269**:1112–15.

Lachenbruch, P. A. (1975). *Discriminant Analysis*. New York: Hafner Press.

Langenheim, J. H. (1995). Early history and progress of women ecologists: emphasis upon research contributions. *Annual Review of Ecology and Systematics* **27**:1–53.

Lawton, J. H. (1995). Ecological experiments with model systems. *Science* **269**:328–31.

Lawton, J. H. (1997). The science and non-science of conservation biology. *Oikos* **79**:3–5.

Lebreton, J. D., Burnham, K. P. Clobert, J., and Anderson, D. R. (1992). Modeling survival and testing biological hypotheses using marked animals: a unified approach with case studies. *Ecological Monographs* **62**:67–118.

Lélé, S., and Norgaard, R. B. (1996). Sustainability and the scientist's burden. *Conservation Biology* **10**:354–65.

Leopold, A. (1924). Grass, brush, timber, and fire in southern Arizona. *Journal of Forestry* **22**:1–10.

Levin, S. A. (1992). The problem of pattern and scale in ecology. *Ecology* **73**:1943–67.

Levy, E. B., and Madden, E. A. (1933). The point method for pasture analysis. *New Zealand Journal of Agriculture* **46**:267–79.

Likens, G. E. (1985). An experimental approach for the study of ecosystems. *Journal of Ecology* **73**:381–96.

Likens, G. E. (ed.) (1989). *Long-Term Studies in Ecology*. Berlin: Springer-Verlag.

Likens, G. E. (1998). Limitations to intellectual progress in ecosystem science. In *Successes, Limitations, and Frontiers in Ecosystem Science*, eds. M. Pace and P. Groffman, pp. 247–71. New York: Springer-Verlag.

Lindström, E. R., Brainerd, S. M., Helldin, J. O., and Overskaug, K. (1995). Pine marten-red fox interactions: a case of interguild predation? *Annales Zoologici Fennici* **32**:123–30.

Litvaitis, J. A. (2001). Importance of early successional habitats to mammals in eastern forests. *Wildlife Society Bulletin* **29**:466–73.

Lloyd, J., Mannan, R. W., DeStefano, S., and Kirkpatrick, C. (1998). The effects of mesquite invasion on a southeastern Arizona grassland bird community. *Wilson Bulletin* **110**:403–8.

Longood, R., and Simmel, A. (1972). Organizational resistance to innovation suggested by research. In *Evaluating Action Programs*, ed. C. Weiss, pp. 311–17. Boston, MA: Allyn and Bacon.

Lubchenco, J., Olson, A. M., Brubaker, L. R. *et al.* (1991). The Sustainable Biosphere Initiative: an ecological research agenda. *Ecology* **72**:371–412.

Ludwig, J. A., and Reynolds, J. F. (1988). *Statistical Ecology: A Primer on Methods and Computing*. New York: Wiley Interscience.

Luken, J. O. (1990). *Directing Ecological Succession*. New York: Chapman and Hall.

MacArthur, R. H. (1957). On the relative abundance of bird species. *Proceedings of the National Academy of Sciences USA* **43**:293–5.

MacNally, R. C. (1983). On assessing the significance of interspecific competition to guild structure. *Ecology* **64**:1646–52.

Madison Botanical Congress. (1894). Congress Proceedings. Madison, Wisconsin, 23–24 August 1893.

Magurran, A. E. (1988). *Ecological Diversity and Its Measurement*. Princeton, NJ: Princeton University Press.

Mahoney, M. J. (1976). *Scientist as Subject: The Psychological Imperative*. Cambridge, MA: Ballinger Press.

Major, J. (1951). A functional factorial approach to plant ecology. *Ecology* **32**:392–412.

Manly, B. F. J. (1986). *Multivariate Statistical Methods: A Primer*. New York: Chapman and Hall.

Margalef, R. (1951). La teoria de la infomacion en ecologia. *Memorias Real Academy Gencia Artes Barcelona* **32**:373–449.

Matter, W. J., and Mannan, R. W. (1989). More on gaining reliable knowledge: a comment. *Journal of Wildlife Management* **53**:1172–6.

May, R. M. (1975). Patterns of species abundance and diversity. In *Ecology and Evolution of Communities*, eds. M. L. and J. M. Diamond, pp. 81–120. Cambridge, MA: Belknap Press, Harvard University Press.

McAuliffe, J. R. (1988). Markovian dynamics of simple and complex desert plant communities. *American Naturalist* **131**:459–90.

McClanahan, T. R., and Wolfe, R. W. (1993). Accelerating forest succession in a fragmented landscape: the role of birds and perches. *Conservation Biology* **7**:279–88.

McClaran, M. P., and McPherson, G. R. (1995). Can soil organic carbon isotopes be used to describe grass-tree dynamics at a savanna-grassland ecotone and within the savanna? *Journal of Vegetation Science* **6**:857–62.

McCune, B. (1997). Influence of noisy environmental data on canonical correspondence analysis. *Ecology* **78**:2617–23.

McIntosh, R. P. (1967). An index of diversity and the relation of certain concepts to diversity. *Ecology* **48**:392–404.

McIntosh, R. P. (1980). The relationship between succession and the recovery process in ecosystems. In *The Recovery Process in Damaged Ecosystems*, ed. J. Cairns, pp. 11–62. Ann Arbor, MI: Ann Arbor Science Publishers.

McPherson, G. R. (1997). *Ecology and Management of North American Savannas.* Tucson, AZ: University of Arizona Press.

McPherson, G. R. (2001a). Teaching and learning the scientific method. *American Biology Teacher* **63**:242–5.

McPherson, G. R. (2001b). Invasive plants and fire: integrating science and management: the need for integration. *Tall Timbers Research Station Miscellaneous Publication* **11**:141–6.

McPherson, G. R., and Weltzin, J. F. (2000). Disturbance and climate change in United States/Mexico borderland plant communities: a state-of-the-knowledge review. USDA Forest Service Rocky Mountain Research Station General Technical Report RMRS-GTR-50, Fort Collins, CO.

McPherson, G. R., Boutton, T. W., and Midwood, A. J. (1993). Stable carbon isotope analysis of soil organic matter illustrates vegetation change at the grassland/woodland boundary in southeastern Arizona, USA. *Oecologia* **93**:95–101.

McPherson, G. R., Rasmussen, G. A., Wester, D. B., and Masters, R. A. (1991). Vegetation and soil zonation associated with *Juniper pinchotii* Sudw. trees. *Great Basin Naturalist* **51**:316–24.

Medawar, P. (1984). *Pluto's Republic.* New York: Oxford University Press.

Mehlert, S., and McPherson, G. R. 1996. Effects of basal area or density as sampling metrics on oak woodland cluster analyses. *Canadian Journal of Forest Research* **26**:38–44.

Menard, S. (1995). *Applied Logistic Regression Analysis.* Thousand Oaks, CA: Sage Publications.

Miles, J. (1979). *Vegetation Dynamics.* London: Chapman and Hall.

Milligan, G. W., and Cooper, M. C. (1985). An examination of procedures for determining the number of clusters in a data set. *Psychometrika* **50**:159–79.

Minchin, P. R. (1987). An evaluation of the relative robustness of techniques for ecological ordination. *Vegetatio* **69**:89–107.

Mobius, K. (1877). *Die Auster and die Austerwirtschaft.* Berlin: Wiegumdt, Hempel and Parey.

Molinari, J. (1989). A calibrated index for the measurement of evenness. *Oikos* **56**:319–26.

Moore, D. R. J., and Keddy, P. A. (1989). The relationships between species richness and standing crop in wetlands: the importance of scale. *Vegetatio* **79**:99–106.

Moore, D. R. J., Keddy, P. A., Gaudet, C. L., and Wisheu, I. C. (1989). Conservation of wetlands: do infertile wetlands deserve a higher priority? *Biological Conservation* **47**:203–17.

Morrison, M. L., Marcot, B., and Mannan, R. W. (1998). *Wildlife-Habitat Relationships*, 2nd edn. Madison, WI: University of Wisconsin Press.

Motomura, I. (1947). Further notes on the law of geometrical progression of the population density in animal association. *Physiological Ecology* **1**:55–60.

Mueller-Dombois, D., and Ellenberg, H. (1974). *Aims and Methods of Vegetation Ecology*. New York: Wiley.

Mumme, R. L., Koenig, W. D., and Pitelka, F. A. (1983). Are acorn woodpecker territories aggregated? *Ecology* **64**:1305–7.

Nee, S., Harvey, P. H., and Cotgreave, P. (1992). Population persistence and the natural relationship between body size and abundance. In *Conservation of Biodiversity for Sustainable Development*, eds. O. T. Sandlund, K. Hindar and A. H. D. Brown, pp. 124–36. Olso, Sweden: Scandinavian University Press.

Neilson, R. P. (1986). High resolution climatic analysis and Southwest biogeography. *Science* **232**:27–34.

Neilson, R. P., and Wullstein, L. H. (1983). Biogeography of two southwest American oaks in relation to atmospheric dynamics. *Journal of Biogeography* **10**:275–97.

Newton, I. (1991). Habitat variation and population regulation in sparrowhawks. *Ibis* **133**:76–88.

Newton, I. (1993). Age and site fidelity in female sparrowhawks, *Accipiter nisus*. *Animal Behaviour* **46**:161–8.

Noble, I. R., and Slatyer, R. O. (1980). The use of vital attributes to predict successional changes in plant communities subject to recurrent disturbances. *Vegetatio* **43**:5–21.

O'Donoghue, M., Boutin, S., Krebs, C. J., and Hofer, E. J. (1997). Numerical responses of coyotes and lynx to the snowshoe hare cycle. *Oikos* **80**:150–62.

O'Donoghue, M., Boutin, S., Krebs, C. J., Zuleta, G., Murphy, D. L., and Hofer, E. J. (1998). Functional responses of coyotes and lynx to the snowshoe hare cycle. *Ecology* **79**:1193–208.

Odum, E. P. (1983). *Basic Ecology*. Philadelphia, PA: Saunders College Publishing.

Odum, H. T., Cantlon, J. E. and Kornicker, L. S. (1960). An organization hierarchy postulate for the interpretation of species–individual distributions, species entropy, ecosystem evolution and the meaning of a species–variety index. *Ecology* **41**:395–9.

Økland, R. H. (1996). Are ordination and constrained ordination alternative or complementary strategies in general ecological studies? *Journal of Vegetation Science* **7**:289–92.

Olson, M. H., Mittelbach, G. G., and Osenberg, C. W. (1995). Competition between predator and prey: resource-based mechanisms and implications for stage-structured dynamics. *Ecology* **76**:1758–71.

Orlóci, L. (1978). *Multivariate Analysis in Vegetation Research*, 2nd edn. The Hague, The Netherlands: Junk.

Owen-Smith, N. (1989). Morphological factors and their consequences for resource partitioning among African savanna ungulates: a simulation modelling approach. In *Patterns in the Structure of Mammalian Communities*, eds. D. W. Morris, Z. Abramsky, B. J. Fox and M. R. Willig, pp. 155–65. Lubbock, TX: Texas Tech University Press.

Owen-Smith, N., and Novellie, P. (1982). What should a clever ungulate eat? *American Naturalist* **119**:151–78.

Paine, R. T. (1963). Trophic relationships of eight sympatric predatory gastropods. *Ecology* **44**:63–73.

Palmer, M. W. (1993). Putting things in even better order: the advantages of canonical correspondence analysis. *Ecology* **74**:2215–30.

Parker, M. A., and Root, R. B. (1981). Insect herbivores limit habitat distribution of a native composite *Machaeranthera canescens*. *Ecology* **62**:1390–2.

Parmenter, R. R., and MacMahon, J. A. (1988). Factors influencing species composition and population sizes in a ground beetle community (Carabidae): predation by rodents. *Oikos* **52**:350–6.

Parzen, E., Tanabe, K., and Kitagawa, G. (eds.). (1998). *Selected Papers of Hirotugu Akaikie*. New York: Springer-Verlag.

Pearson, K. (1901). On lines and planes of closest fit to systems of points in space. *Philosophical Magazine, Sixth Series* **2**:559–72.

Peet, R. K. (1974). The measurement of species diversity. *Annual Review of Ecology and Systematics* **5**:285–307.

Peet, R. K., Knox, R. G., Case, J. S., and Allen, R. B. (1988). Putting things in order: the advantages of detrended correspondence analysis. *American Naturalist* **131**:924–34.

Pentecost, A. (1980). Aspects of competition in saxicolous lichen communities. *Lichenologist* **12**:135–44.

Peters, D. P., and Ceci, S. J. (1982). Resubmitting previously published articles: a study of the journal review process in psychology. *Behavioral and Brain Sciences* **5**:187–95.

Peterson, D. L., and Parker, V. T. (eds.) (1998). *Ecological Scale: Theory and Applications*. New York: Columbia University Press.

Peters, R. H. (1980). Useful concepts for predictive ecology. *Synthese* **43**:257–69.

Peters, R. H. (1991). *A Critique for Ecology*. Cambridge: Cambridge University Press.

Petren, K., and Case, T. J. (1996). An experimental demonstration of exploitation competition in an ongoing invasion. *Ecology* **77**:118–32.

Pickett, S. T. A. (1976). Succession: an evolutionary interpretation. *American Naturalist* **110**:107–19.

Pickett, S. T. A., and White, P. S. (1985). *The Ecology of Natural Disturbance and Patch Dynamics*. New York: Academic Press.

Pickett, S. T. A., Collins, S. L., and Armesto, J. J. (1987). A hierarchical consideration of causes and mechanisms of succession. *Vegetatio* **69**:109–14.

Pickett, S. T. A., Kolasa, J., and Jones, C. G. (1994). *Ecological Understanding: The Nature of Theory and the Theory of Nature*. San Diego, CA: Academic Press.

Pielou, E. C. (1966). The measurement of diversity in different types of biological collections. *Journal of Theoretical Biology* **13**:131–44.

Pielou, E. C. (1975). *Ecological Diversity*. New York: Wiley.

Pielou, E. C. (1977). *Mathematical Ecology*. New York: Wiley.

Pielou, E. C. (1979). On A. J. Underwood's model for a random pattern. *Oecologia* **44**:143–4.

Pielou, E. C. (1981). The usefulness of ecological models: a stock-taking. *Quarterly Review of Biology* **56**:17–31.

Pielou, E. C. (1984). *The Interpretation of Ecological Data: A Primer on Classification and Ordination*. New York: Wiley.

Platt, J. R. (1964). Strong inference. *Science* **146**:347–53.

Podani, J. (1984). Spatial processes in the analysis of vegetation: theory and review. *Acta Botanica Hungarica* **30**:75–118.

Polis, G. A., and McCormick, S. J. (1986). Scorpions, spiders, and solpugids: predation and competition among distantly related taxa. *Oecologia* **71**:111–16.

Polis, G. A., Myers, C. A., and Holt, R. D. (1989). The evolution and ecology of intraguild predation: competitors that eat each other. *Annual Review of Ecology and Systematics* **20**:297–330.

Pollock, K. H., Nichols, J. D., Brownie, C., and Hines, J. E. (1990). Statistical inference for capture–recapture experiments. *Wildlife Monographs* **107**:1–97.

Popper, K. (1959). *The Logic of Scientific Discovery*. New York: Basic Books.

Popper, K. (1981). Science, pseudo-science, and falsifiability. In *On Scientific Thinking*, eds. R. D. Tweney, M. E. Doherty and C. R. Mynatt, pp. 92–9. New York: Columbia University Press.

Prentice, I. C. (1986). Vegetation responses to past climatic variation. *Vegetatio* **67**:131–41.

Prentice, I. C., and van der Maarel, E. (1987). *Theory and Models in Vegetation Science*. Reprinted from Vegetatio 69. Dordrecht, The Netherlands: Junk.

Preston, F. W. (1948). The commonness, and rarity, of species. *Ecology* **29**:254–83.

Prodgers, R. A. (1984). Collection and analysis of baseline vegetation data. *Minerals and the Environment* **6**:101–4.

Ralph, C. J., and Scott, J. M. (1981). Estimating numbers of terrestrial birds. *Studies in Avian Biology* **6**:1–630.

Ramensky, L. G. (1930). Zur Methodik der vergleichenden Bearbeitung und Ordnung von Pflanzenlisten und anderen Objekten, die durch mehrere, verschiedenartig wirkende Faktoren bestimmt werden. *Bieträge zur Biologie der Pflanzen* **18**:269–304.

Ramsey, F. L., and Schafer, D. W. (1997). *The Statistical Sleuth: A Course in Methods of Data Analysis*. Belmont, CA: Duxbury Press.

Rexstad, E. A., Miller, D. D., Flather, C. H., Anderson, E. M., Hupp, J. W., and Anderson, D. R. (1988). Questionable multivariate statistical inference in wildlife habitat and community studies. *Journal of Wildlife Management* **52**:794–8.

Risser, P. G. (ed.) (1991). *Long-Term Ecological Research: An International Perspective*. New York: Wiley.

Robinson, G. R., and Handel, S. N. (1993). Forest restoration on a closed landfill: rapid addition of new species by bird dispersal. *Conservation Biology* **7**:271–8.

Romesburg, H. C. (1981). Wildlife science: gaining reliable knowledge. *Journal of Wildlife Management* **45**:293–313.

Romesburg, H. C. (1984). *Cluster Analysis*. Belmont, CA: Lifetime Learning Publications.

Rosenzweig, M. L., and Abramsky, A. (1986). Centrifugal community organization. *Oikos* **46**:339–45.

Rusch, D. H., Meslow, E. C., Doerr, P. D., and Keith, L. B. (1972). Response of great horned owl populations to changing prey densities. *Journal of Wildlife Management* **36**:282–96.

Sackville Hamilton, N. R. (1994). Replacement and additive designs for plant competition studies. *Journal of Applied Ecology* **31**:599–603.

Scholes, R. J., and Archer, S. R. (1997). Tree–grass interactions in savannas. *Annual Review of Ecology and Systematics* **28**:517–44.

Seber, G. A. F. (1982). *The Estimation of Animal Abundance and Related Parameters*, 2nd edn. New York: Macmillan.

Shannon, C. E., and Weaver, W. (1949). *The Mathematical Theory of Communication*. Urbana, IL: University of Illinois Press.

Sharitz, R. R., Boring, L. R., Van Lear, D. H., Pinder, J. E., III. (1992). Integrating ecological concepts with natural resource management of southern forests. *Ecological Applications* **2**:226–37.

Shipley, B., and Keddy, P. A. (1987). The individualistic and community-unit concepts as falsifiable hypotheses. *Vegetatio* **69**:47–55.

Shreve, F. (1915). *The Vegetation of a Desert Mountain Range as Conditioned by Climatic Factors*. Washington, DC: Carnegie Institute of Washington Publication 217.

Shreve, F. (1922). Conditions indirectly affecting vertical distribution on desert mountains. *Ecology* 3:269–74.

Simberloff, D. (1983). Competition theory, hypothesis-testing, and other community ecology buzzwords. *American Naturalist* 122:626–35.

Simberloff, D. (1984). The great god of competition. *The Sciences* 24:17–22.

Simpson, E. H. (1949). Measurement of diversity. *Nature* 163:688.

Sinclair, A. R. E. (1975). The resource limitation of trophic levels in tropical grassland ecosystems. *Journal of Animal Ecology* 44:497–520.

Sinclair, A. R. E. (1977). *The African Buffalo: A Study of Resource Limitation of Populations*. Chicago: University of Chicago Press.

Sinclair, A. R. E., Krebs, C. J., and Smith, J. N. M. (1982). Diet quality and food limitation in herbivores: the case of the snowshoe hare. *Canadian Journal of Zoology* 60:889–97.

Smallwood, K. S. (1995). Scaling Swainson's hawk population density for assessing habitat use across an agricultural landscape. *Journal of Raptor Research* 29:172–8.

Smallwood, K. S., and Schonewald, C. (1996). Scaling population density and spatial pattern for terrestrial, mammalian carnivores. *Oecologia* 105:329–35.

Smartt, P. F. M., Meacock, S. E. and Lambert, J. M. (1974). Investigations into the properties of quantitative vegetation data. I. Pilot study. *Journal of Ecology* 62:735–59.

Snaydon, R. W. (1991). Replacement or additive designs for competition studies? *Journal of Applied Ecology* 28:930–46.

Snaydon, R. W. (1994). Replacement and additive designs revisited: comments on the review paper by N. R. Sackville Hamilton. *Journal of Applied Ecology* 31:784–6.

Sneath, P. H. A., and Sokal, R. R. (1973). *Numerical Taxonomy*. San Francisco, CA: W H Freeman.

Sokal, R. R., and Michener, C. D. (1958). A statistical method for evaluating systematic relationships. *University of Kansas Science Bulletin* 38:1409–38.

Spiller, D. A., and Schoener, T. W. (1998). Lizards reduce spider species richness by excluding rare species. *Ecology* 79:503–16.

Spurr, S. H. (1952). Origin of the concept of forest succession. *Ecology* 33:426–7.

Stanley, T. R., Jr. (1995). Ecosystem management and the arrogance of humanism. *Conservation Biology* 9:255–62.

Stapp, P. (1997). Community structure of shortgrass-prairie rodents: competition or risk of intraguild predation? *Ecology* 78:1519–30.

Tansley, A. G. (1914). Presidential address. *Journal of Ecology* 2:194–202.

Taylor, J. (1990). Questionable multivariate statistical inference in wildlife habitat and community studies: a comment. *Journal of Wildlife Management* 54:186–9.

ter Braak, C. J. F. (1986). Canonical correspondence analysis: a new eigenvector technique for multivariate direct gradient analysis. *Ecology* 67:1167–79.

ter Braak, C. J. F. (1988). CANOCO. Technical Report LWA-88-02, Agricultural Mathematics Group. Wageningen, The Netherlands.

ter Braak, C. J. F., and Prentice, I. C. (1988). A theory of gradient analysis. *Advances in Ecological Research* 18:271–313.

Thompson, F. R., III, and DeGraaf, R. M. (2001). Conservation approaches for woody, early successional communities in the eastern United States. *Wildlife Society Bulletin* 29:483–94.

Thompson, W. L., White, G. C., and Gowan, C. (1998). *Monitoring Vertebrate Populations*. San Diego, CA: Academic Press.

Tilman, D. (1982). *Resource Competition and Community Structure*. Princeton, NJ: Princeton University Press.

Tilman, D. (1985). The resource-ratio hypothesis of plant succession. *American Naturalist* **125**:827–52.

Tilman, D. (1988). *Plant Strategies and the Dynamics and Structure of Plant Communities*. Princeton, NJ: Princeton University Press.

Tilman, D. (1990). Constraints and trade-offs: toward a predictive theory of competition and succession. *Oikos* **58**:3–15.

Tilman, D., and Wedin, D. (1991). Oscillations and chaos in the dynamics of a perennial grass. *Nature* **353**:653–5.

Todd, A. W., Keith, L. B., and Fischer, C. A. (1981). Population ecology of coyotes during a cyclic fluctuation of snowshoe hares. *Journal of Wildlife Management* **45**:629–40.

Trani, M. K., Brooks, R. T., Schmidt, T. L., Rudis, V. A., and Gabbard, C. M. (2001). Patterns and trends of early successional forests in the eastern United States. *Wildlife Society Bulletin* **29**:413–24.

Turkington, R., and Jolliffe, P. A. (1996). Interference in *Trifolium repens–Lolium perenne* mixtures: short- and long-term relationships. *Journal of Ecology* **84**:563–71.

Underwood, A. J. (1978). The detection of non-random patterns of distribution of species along an environmental gradient. *Oecologia* **36**:317–26.

Underwood, A. J. (1995). Ecological research and (and research into) environmental management. *Ecological Applications* **5**:232–47.

Vallentine, J. F. (1990). *Grazing Management*. San Diego, CA: Academic Press.

Van Auken, O. W., and Bush, J. K. (1997). Growth of *Prosopis glandulosa* in response to changes in aboveground and belowground interference. *Ecology* **78**:1222–9.

van der Maarel, E. (1979). Transformation of cover-abundance values in phytosociology and its effects on community similarity. *Vegetatio* **39**:97–114.

van Groenewoud, H. (1992). The robustness of correspondence analysis, detrended correspondence analysis, and TWINSPAN analysis. *Journal of Vegetation Science* **3**:239–46.

van Horne, B. (1983). Density as a misleading indicator of habitat quality. *Journal of Wildlife Management* **47**:893–901.

Vaughan, M. R., and Keith, L. B. (1981). Demographic response of experimental snowshoe hare populations to overwinter food shortage. *Journal of Wildlife Managment* **45**:354–80.

Vickery, P. D., Hunter, M. L., Jr., and Wells, J. V. (1992a). Is density an indicator of breeding success? *Auk* **109**:706–10.

Vickery, P. D., Hunter, M. L., Jr., and Wells, J. V. (1992b). Use of a new reproductive index to evaluate relationships between habitat quality and breeding success. *Auk* **109**:697–705.

Vitousek, P. M., Ehrlich, P. R., Ehrlich, A. H., and Matson, P. A. (1986). Human appropriation of the products of photosynthesis. *BioScience* **36**:368–73.

Ward, D. A., and Kassebaum, G. S. (1972). On biting the hand that feeds: some implications of correctional effectiveness. In *Evaluating Action Programs*, ed. C. Weiss, pp. 300–10. Boston, MA: Allyn and Bacon.

Ward, J. H. (1963). Hierarchical grouping to optimize an objective function. *Journal of the American Statistical Association* **58**:236–44.

Ward, R. M. P., and Krebs, C. J. (1985). Behavioural responses of lynx to declining snowshoe hare abundance. *Canadian Journal of Zoology* **63**:2817–24.

Wartenberg, D., Ferson, S., and Rohlf, F. J. (1987). Putting things in order: a critique of detrended correspondence analysis. *American Naturalist* **129**:434–48.

Watkins A. J., and Wilson, J. B. (1994). Plant community structure, and its relation to the vertical complexity of communities: dominance/diversity and spatial rank consistency. *Oikos* **70**:91–8.

Weiner, J. (1990). Asymmetric competition in plant populations. *Trends in Ecology and Evolution* **5**:360–4.

Weiner, J. (1995). On the practice of ecology. *Journal of Ecology* **83**:153–8.

Weltzin, J. K., Archer, S. R., and Heitschmidt, R. K. (1997). Small-mammal regulation of vegetation structure in a temperate savanna. *Ecology* **78**:751–63.

Werger, M. J. A., Louppen, J. M. W., and Eppink, J. H. M. (1983). Species performance and vegetation boundaries along an environmental gradient. *Vegetatio* **52**:141–50.

Werner, P. A. (1977). Colonization success of a "biennial" plant species: experimental field studies of species cohabitation and replacement. *Ecology* **58**:840–9.

West, N. E., and Reese, G. A. (1991). Comparison of some methods for collecting and analyzing data on aboveground net production and diversity of herbaceous vegetation in a northern Utah subalpine context. *Vegetatio* **96**:145–63.

Wester, D. B., and Wright, H. A. (1987). Ordination of vegetation change in Guadalupe Mountains, New Mexico, USA. *Vegetatio* **72**:27–33.

Westoby, M., Walker, B., and Noy-Meir, I. (1989). Opportunistic management for rangelands not at equilibrium. *Journal of Range Management* **42**:266–74.

Whitham, T. G., Maschinski, J., Larson, K. C., and Paige, K. N. (1991). Plant responses to herbivory: the continuum from negative to positive and underlying physiological mechanisms. In *Plant-Animal Interactions: Evolutionary Ecology in Tropical and Temperate Regions*, eds. P. W. Price, T. M. Lewinsohn, G. W. Fernandes and W. W. Benson, pp. 227–56. New York: Wiley.

Whittaker, R. H. (1956). Vegetation of the Great Smoky Mountains. *Ecological Monographs* **26**:1–80.

Whittaker, R. H. (1960). Vegetation of the Siskiyou Mountains, Oregon and California. *Ecological Monographs* **30**:279–338.

Whittaker, R. H. (1962). Classification of natural communities. *Botanical Review* **28**:1–239.

Whittaker, R. H. (1965). Dominance and diversity in land plant communities. *Science* **147**:250–60.

Whittaker, R. H. (1967). Gradient analysis of vegetation. *Biological Review* **42**:207–64.

Whittaker, R. H. (1972). Evolution and measurement of species diversity. *Taxon* **21**:213–51.

Whittaker, R. H. (1975). *Communities and Ecosystems*, 2nd edn. New York: Macmillan.

Whittaker, R. H., and Niering, W. A. (1965). Vegetation of the Santa Catalina Mountains, Arizona: a gradient analysis of the south slope. *Ecology* **46**:429–52.

Whittaker, R. H., and Niering, W. A. (1975). Vegetation of the Santa Catalina Mountains, Arizona. V. Biomass, production, and diversity along the elevation gradient. *Ecology* **56**:771–90.

Wicklum, D., and Davies, R. W. (1995). Ecosystem health and integrity? *Canadian Journal of Botany* **73**:997–1000.

Williams, B. K. (1983). Some observations on the use of discriminant analysis in ecology. *Ecology* **64**:1283–91.

Williams, D. G., McPherson, G. R., and Weltzin, J. F. (1999). Stress in wildland plants: implications for ecosystem structure and function. In *Handbook of Plant and Crop Stress*, revised and expanded edition, ed. M. Pessarakli, pp. 907–29. New York: Marcel Dekker.

Willson, M. F. (1981). Commentary: ecology and science. *Bulletin of the Ecological Society of America* **62**:4–12.

Wilson, D. S. (1997). Biological communities as functionally organized units. *Ecology* **78**:2018–24.

Wilson, D. S., and Sober, E. (1989). Reviving the superorganism. *Journal of Theoretical Biology* **136**:337–56.

Wilson, J. B. (1991). Methods for fitting dominance/diversity curves. *Journal of Vegetation Science* **2**:35–46.

Wilson, J. B. (1993). Would we recognize a broken-stick community if we found one? *Oikos* **67**:181–3.

Wilson, J. B., Wells, T. C. E., Trueman, I. C. *et al.* (1996). Are there assembly rules for plant species abundance? An investigation in relation to soil resources and successional trends. *Journal of Ecology* **84**:527–38.

Wisheu, I. C., and Keddy, P. A. (1992). Competition and centrifugal organization of plant communities: theory and tests. *Journal of Vegetation Science* **3**:147–56.

Witty, J. E., and Knox, E. G. (1964). Grass opal in some chestnut and forested soils in north-central Oregon. *Soil Science Society of America Proceedings* **28**:685–8.

Wondzell, S., and Ludwig, J. A. (1995). Community dynamics of desert grasslands: influences of climate, landforms, and soils. *Journal of Vegetation Science* **6**:377–90.

Wright, S. J., and Biehl, C. C. (1982). Island biogeographic distributions: testing for random, regular, and aggregated patterns of species occurrence. *American Naturalist* **119**:345–57.

Yeaton, R. I., Travis, J., and Gilinsky, E. (1977). Competition and spacing in plant communities: the Arizona upland association. *Journal of Ecology* **65**:587–95.

Young, C. C. (1998). Defining the range: the development of carrying capacity in management practice. *Journal of the History of Biology* **31**:61–83.

Index

Note: page numbers in italics refer to boxes; those in bold refer to illustrations.